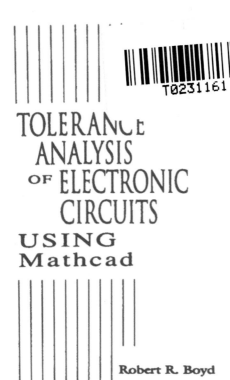

TOLERANCE ANALYSIS OF ELECTRONIC CIRCUITS
USING Mathcad

Robert R. Boyd

CRC Press
Taylor & Francis Group
Boca Raton London New York

CRC Press is an imprint of the
Taylor & Francis Group, an **informa** business

CRC Press
Taylor & Francis Group
6000 Broken Sound Parkway NW, Suite 300
Boca Raton, FL 33487-2742

©1997 by Taylor & Francis Group, LLC
CRC Press is an imprint of Taylor & Francis Group, an Informa business

Visit the Taylor & Francis Web site at
http://www.taylorandfrancis.com

and the CRC Press Web site at
http://www.crcpress.com

Preface

This book is written for the practicing electronics professional. Knowledge of the capabilities and limitations of tolerance analysis is a valuable asset to both the engineer and senior technician.

Tolerance analysis is necessary in several phases of the design task, primarily to show that a circuit card, and a system of circuit cards, will meet requirements over production life. Methods are shown which can be used in the design process to perform worst-case analysis, determine manufacturing yields, calculate limits for production testing, determine if a design meets specification limits, and for component stress analysis.

Topics include extreme value analysis and root-sum-square analysis using symmetric and asymmetric tolerances, Monte Carlo analysis using normal and uniform component distributions, Spice comparisons, sensitivity formulas, and ratiometric tolerances. Also included are tolerance analyses of opamp offsets and anomalies of high-Q and high-gain circuits.

Much of this material is not found in textbooks prior to 1999. The information will benefit those involved in the following areas of engineering: analog design/analysis, RF design/analysis, test, system, reliability, and quality assurance. The book will also be of value to specification writers, programmers, and senior technicians.

About the Author

Robert R. Boyd was a technical instructor in the United States Air Force for 19 years. Upon his retirement from the USAF in 1971, he enrolled at the University of New Mexico and obtained a BSEE degree with honors in 1974. Mr. Boyd was subsequently employed in the aerospace industry in analog circuit design until 1996. He is presently a consultant in analog circuit design and analysis and teaches a course in the tolerance analysis of electronic circuits at the University of California Extension (Irvine, CA).

Contents

PART 3
Advanced Topics 115

Introduction

Some tolerance analysis methods are given in outline form below:

Worst-case analysis
- Extreme Value Analysis (EVA) using sensitivity signs
- EVA using all tolerance combinations (Fast Monte Carlo Analysis) (FMCA)

Non-worst-case analysis
- Monte Carlo Analysis (MCA)
- Root-Sum-Square (RSS) analysis

The treatment of asymmetric tolerances is neglected in what few existing publications there are on the subject prior to 1999 (see References). Asymmetric tolerances can occur when performing design verification analyses where the specification temperature extremes, and hence the temperature variation of the components, are most always asymmetric. For example, –55°C to +70°C. The worst-case methods used in circuit analysis packages derived from Spice cannot accommodate asymmetric tolerances. In addition, for ac circuits, the sensitivity portion and the resulting output of the Spice. WCASE analysis is incorrect as will be shown.

In general, dc tolerance analysis is straightforward and yields no surprises using the four methods given above. In ac analysis, especially in moderate to high Q filter circuits, there are some pitfalls the analyst must be aware of. These anomalies are illustrated by several examples.

The formulas and equations were created using Mathcad 7.0. Some familiarity with Mathcad is assumed.

This material is based on class notes for a course developed by the author for the University of California Extension, Irvine, CA.

1 Root-Sum-Square (RSS) and Extreme Value Analysis (EVA)

MATHCAD DERIVATIVES

Derivatives are used extensively in tolerance analysis. The following example shows how Mathcad takes the derivative of a function. A simple voltage divider is used as the example function.

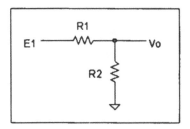

First, we assign values to the input voltage E1, and the two resistors R1 and R2: R1 := 1000, R2 := 3000, E1 :=10. If we want to dress this up a little bit, we can use unit suffixes:

$$K := 10^3, V := 1, R1 := 1 \cdot K, R2 := 3 \cdot K, E1 := 10 \cdot V$$

Next, we create a Mathcad function that expresses the voltage divider equation. These functions are much like subroutines in programming languages such as Fortran or QBasic. We can then "call" them and pass original or modified parameters to get the answers we need.

Now we call this function and assign the nominal value to variable Vo:

$$Vo := G(R1, R2, E1) \quad Vo = 7.5$$

If we wanted to change the value of R2, we could multiply it by, say, 1.1, and then call the function:

$$Vo1 := G(R1, R2 \cdot 1.1, E1) \quad Vo1 = 7.674$$

We now take the derivative with respect to R1 and assign it to a variable name we create called "dmR1", for derivative, Mathcad, R1.

$$dmR1 := \frac{d}{dR1} G(R1, R2, E1) \quad dmR1 = -1.875 \cdot 10^{-3}$$

Next, the derivative with respect to (wrt) R2:

$$dmR2 := \frac{d}{dR2} G(R1, R2, E1) \quad dmR2 = 6.25 \cdot 10^{-4}$$

And finally, the derivative wrt input voltage E1:

$$dmE1 := \frac{d}{dE1} G(R1, R2, E1) \quad dmE1 = 0.75$$

For sake of brevity, we would like to take all the derivatives at once, i.e., a one-line iteration. First, we create a column vector of the three components:

$$Mc = (R1 \; R2 \; E1)^T \qquad Mc^T = [1000 \; 3000 \; 10]$$

We create a counter variable to reference Mc and call it p: $p := 1..3$.

Now Mc_1 should then pick out R1, Mc_2, R2 and so forth: $Mc_1 = 1000$, $Mc_2 = 3000$, $Mc_3 = 10$. We attempt to create the three derivatives in one line as follows:

$$dm_p := \frac{d}{dMc_p} G(R1,R2,E1)$$

Mathcad flags this as an error and will not perform the operation. Hence we must resort to creating our own derivatives that can be iterated.

APPROXIMATE DERIVATIVES

We perturbate one component, i.e., change its value by a very small amount:

dpf := 0.0001   (dpf is called the derivative purturbation factor)
Vrl := $G(R1\cdot(1 + dpf),R2,E1)$ Vrl = 7.4998

Vrl corresponds to the familiar calculus expression $y(x) = f(x + \Delta x)$. The derivative is defined as:

$$\frac{dy}{dx} = \frac{f(x + \Delta x) - f(x)}{\Delta x}$$

We can create this function in Mathcad and call it daR1, for derivative, approximate, R1:

$$daR1 := \frac{Vrl - Vo}{dpf \cdot R1} \quad daR1 = -1.875 \cdot 10^{-3}$$

How close is this to dmR1?

$$dmR1 = -1.875 \cdot 10^{-3}$$

We compare the two by performing a relative error check of

$$\frac{\text{true} - \text{measured}}{\text{true}} \quad \text{or:} \quad 1 - \frac{daR1}{dmR1} = 0.002 \cdot \%$$

As will be seen, this is close enough for our purposes.
Normalized sensitivities are defined as:

$$\frac{x}{y} \cdot \frac{dy}{dx}$$

For R1:

$$St1 := \frac{R1 \cdot dmR1}{Vo} \quad \text{or} \quad St1 = -0.25$$

The approximate normalized sensitivity of R1 is then:

$$Sen1 := \frac{R1 \cdot daR1}{Vo} \quad Sen1 := \frac{R1}{Vo} \cdot \frac{Vr1 - Vo}{dpf \cdot R1}$$

$$Sen1 := \left(\frac{Vr1}{Vo} - 1\right) \cdot \frac{1}{dpf} \quad Sen1 = -0.25$$

For R2:

$$Vr2 := G(R1, R2 \cdot (1 + dpf), E1)$$

$$Sen2 := \left(\frac{Vr2}{Vo} - 1\right) \cdot \frac{1}{dpf} \quad Sen2 = 0.25$$

and similarly for E1. We are now in a position to perform an iteration. First, we create a perturbation matrix Q as follows:

$$\text{identity}(3) = \begin{bmatrix} 1 & 0 & 0 \\ 0 & 1 & 0 \\ 0 & 0 & 1 \end{bmatrix}$$

$$\text{dpf} \cdot \text{identity}(3) = \begin{bmatrix} 0.0001 & 0 & 0 \\ 0 & 0.0001 & 0 \\ 0 & 0 & 0.0001 \end{bmatrix}$$

$$Q := \text{dpf} \cdot \text{identity}(3) + 1 \qquad Q = \begin{bmatrix} 1.0001 & 1 & 1 \\ 1 & 1.0001 & 1 \\ 1 & 1 & 1.0001 \end{bmatrix}$$

Using counter p, we perform the three $y(x) = f(x) + \Delta x$ operations as follows:

$$Vr_p := G(R1 \cdot Q_{p,1}, R2 \cdot Q_{p,2}, E1 \cdot Q_{p,3})$$
$$Vr^T = [7.4998 \quad 7.5002 \quad 7.5008]$$

Getting the iterated normalized sensitivities is now easy:

$$Sen_p := \left(\frac{Vr_p}{Vo} - 1 \right) \cdot \frac{1}{dpf} \qquad Sen^T = [-0.25 \quad 0.25 \quad 1]$$

For practical considerations, this is the recommended form. Taking the derivatives and sensitivities of a circuit with many components would be too lengthy and error prone.

EXTREME VALUE ANALYSIS (EVA)

Now that we have the sensitivities, what do we do with them? The sensitivity signs tell us which way the output goes when we change a component value. For example, if we increase R1, the output will decrease since the sensitivity is negative. Conversely, if we decrease R1, the output will increase. It works both ways. This can stated as: if we increase (decrease) R1, the output will decrease (increase). For R2, (plus sensitivity), if we increase (decrease) R2, the output will increase (decrease). After obtaining the tolerances of R1, R2, and E1, we can determine the maximum possible (extreme value) output of the voltage divider as follows: First we assign tolerances in decimal percent: 1% for the resistors and 5% for the input voltage E1.

$$\text{Tr} := 0.01 \qquad \text{Te} := 0.05$$

and create a tolerance array T:

$$T := \begin{bmatrix} -\text{Tr} & -\text{Tr} & -\text{Te} \\ \text{Tr} & \text{Tr} & \text{Te} \end{bmatrix}$$

In the tolerance array, the negative tolerances must always be placed in the first row and the positive tolerances must always be put in the second row.

$$T = \begin{bmatrix} -0.01 & -0.01 & -0.05 \\ 0.01 & 0.01 & 0.05 \end{bmatrix}$$

To get the maximum output (Vevh), we want to decrease R1 since its sign is negative. Since the signs of R2 and E1 are plus, we want to increase their values.

$$\text{Vevh} := G\left(R1 \cdot (1 - \text{Tr}), R2 \cdot (1 + \text{Tr}), E1 \cdot (1 + \text{Te})\right) \qquad \text{Vevh} = 7.914$$

To get the minimum extreme value (Vevl), change the sign of all the foregoing tolerances:

$$\text{Vevl} := G(R1 \cdot (1 + Tr), R2 \cdot (1 - Tr), E1 \cdot (1 - Te)) \quad \text{Vevl} = 7.089$$

To use the tolerance array T and get both maximum and minimum values in one operation, we first create a row counter, or a low-high counter k:

$$k := 1..2$$

Next, we create a new tolerance array in which the row position is *swapped if the sensitivity is negative*. If the sensitivities are plus, we do nothing. We can do this in two separate operations. The sensitivities are repeated for reference:

$$\text{Sen}^T = [-0.25 \quad 0.25 \quad 1]$$

For row 1 of the T array:

$$M_{1,p} := \text{if}(\text{Sen}_p < 0, T_{2,p}, T_{1,p}) \quad M = [0.001 \ -0.01 \ -0.05]$$

Adding row 2:

$$M_{2,p} := \text{if}\left(\text{Sen}_p > 0, T_{2,p}, T_{1,p}\right) \quad M = \begin{bmatrix} 0.01 & -0.01 & -0.05 \\ -0.01 & 0.01 & 0.05 \end{bmatrix}$$

We are going to be multiplying the components by 1 + the tolerance. We do this and create a nested "if" statement:

$$M_{k,p} := \text{if}(k = 1, \text{if}(\text{Sen}_p < 0, 1 + T_{2,p}, 1 + T_{1,p}),$$
$$\text{if}(\text{Sen}_p > 0, 1 + T_{2,p}, 1 + T_{1,p}))$$

$$M = \begin{bmatrix} 1.01 & 0.99 & 0.95 \\ 0.99 & 1.01 & 1.05 \end{bmatrix}$$

Now both high and low extreme values can be calculated using one line. To do this, merely multiply by elements of M in R1, R2, E1 column order:

$$\text{Vev}_k := G\big(R1 \cdot M_{k,1}, R2 \cdot M_{k,2}, E1 \cdot M_{k,3}\big) \qquad \text{Vev} = \begin{bmatrix} 7.089 \\ 7.914 \end{bmatrix}$$

The deltas are:

$$\Delta\text{Vev}_k := \text{Vev}_k - \text{Vo} \qquad \Delta\text{Vev} = \begin{bmatrix} -0.411 \\ 0.414 \end{bmatrix}$$

Note that the deltas are not symmetrical.

EVA SUMMARY

1. Define circuit G-function and get the nominal output

$$G(R1, R2, E1) := \frac{E1 \cdot R2}{R1 + R2}$$

$$\text{Vo} := G(R1, R2, E1) \quad \text{Vo} = 7.5$$

2. Create component counter p, the Q matrix, and get sensitivities $p := 1..3$

$$Q := dpf \cdot identity(3) + 1 \qquad Vr_p := G\left(R1 \cdot Q_{p,1}, R2 \cdot Q_{p,2}, E1 \cdot Q_{p,3}\right)$$

$$Sen_p := \left(\frac{Vr_p}{Vo} - 1\right) \cdot \frac{1}{dpf}$$

3. Create the tolerance array:

$$T := \begin{bmatrix} -Tr & -Tr & -Te \\ Tr & Tr & Te \end{bmatrix} \qquad T = \begin{bmatrix} -0.01 & -0.01 & -0.05 \\ 0.01 & 0.01 & 0.05 \end{bmatrix}$$

4. Create the swapped tolerance array:

$$M_{k,p} := if\left(k = 1, \; if(Sen_p < 0, \; 1 + T_{2,p}, \; 1 + T_{1,p}),\right.$$
$$\left. if(Sen_p > 0, \; 1 + T_{2,p}, \; 1 + T_{1,p})\right)$$

$$M = \begin{bmatrix} 1.01 & 0.99 & 0.95 \\ 0.99 & 1.01 & 1.05 \end{bmatrix}$$

5. Get the extreme values by multiplying the function components by the swapped array M:

$$Vev_k := G\left(R1 \cdot M_{k,1}, R2 \cdot M_{k,2}, E1 \cdot M_{k,3}\right) \qquad Vev = \begin{bmatrix} 7.089 \\ 7.914 \end{bmatrix}$$

ROOT-SUM-SQUARE ANALYSIS (RSS)

RSS methods require knowing some aspects of a statistical distribution known as Gaussian (sometimes called normal) distribution. The general characteristics of this familiar bell-shaped curve are given below:

$$x := -4, -3.8 .. 4 \qquad \sigma := 1 \qquad m := 0$$
(Standard deviation = 1, and mean = 0; normalized.)

$$c := \sqrt{2 \cdot \pi} \qquad y(x) := \frac{1}{\sigma \cdot c} \cdot \exp\left[\frac{-(x-m)^2}{2 \cdot \sigma^2}\right] \qquad u(x) := \int_{-10}^{x} y(x) \, dx$$

$u(x)$ is known as the cumulative normal probability distribution and c is a normalizing constant.

$$u(1) = 0.841 \qquad (\sigma = 1 \text{ is } 84.1\% \text{ of the total } 100\%)$$
$$u(3) = 0.999 \qquad (\sigma = 3 \text{ is } 99.9\% \text{ of the total } 100\%)$$

The 90% point is called the 90th percentile: $xs := 1.2816$ $u(xs) = 0.9$
The 95th percentile: $xv := 1.6449$ $u(xv) = 0.95$
The *variance* is defined as the squared standard deviation: $var = \sigma^2$.

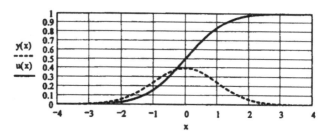

FIGURE 1. Normal distribution plot.

EXAMPLE:

500 college students weigh an average of 151 lbs, with a standard deviation of 15 lbs.

$$s1 := 15 \quad m := 151 \quad N := 500 \quad x := 100, 120 .. 200$$

$$n(x) := \frac{x - m}{s1} \quad \text{(Normalize)}$$

Someone who weighs $m + 3s1 = 196$ weighs as much or more than 99.9% of the students.

90th %tile: $x9 := m + xs \cdot s1 \quad x9 = 170.2$
(More than 90% of students.)

95th %tile: $x95 := m + xv \cdot s1 \quad x95 = 175.67$
(More than 95% of students, etc.)

How many students weigh less than 170 lbs?

$$n(170) = 1.267 \quad u(1.267) = 0.897 \quad 0.897 \cdot N = 448.5$$

Answer:

$$449 \text{ or } N \cdot u(n(170)) = 449$$

What percentage of students weigh more than 160 lbs?

$$1 - u(n\,160)) = 27.43 \cdot \%$$

The standard deviation (σ) is a measure of the spread of the bell shape. A large S indicates a flattened curve while a small s indicates a sharp-peaked curve. For example, let s in the above example be 10 lbs instead of 15.

$$s2 := 10 \quad 3 \cdot s2 + m = 181$$

Now one who weighs 181 lbs weighs more than 99.9% of the students.

MATHCAD BUILT-IN NORMAL
DISTRIBUTION FUNCTIONS

The function pnorm has the syntax $p = pnorm(q, m, sd)$, where m is the mean and sd is the standard deviation. Pnorm in effect goes from the x-axis (q) up to the solid line in the graph (p). Example:

$xs = 1.282$ $p1 := pnorm(xs, 0, 1)$ $p1 = 0.9$ (90th percentile)
$xv = 1.645$ $p2 := pnomr(xv, 0, 1)$ $p2 = 0.95$ (95th percentile)

The function qnorm has the syntax $q = qnorm(p, m, sd)$, and is the inverse of pnorm. It goes from the solid line in the graph (p) down to the x-axis (q).

The 90th and 95th percentiles were found using

$$xs := qnorm(0.9, 0, 1) \qquad xs = 1.2816$$
$$xv := qnorm(0.95, 0, 1) \qquad xv = 1.6449$$

The function dnowm is the probability p of selecting the value x from a normal distribution with mean m and standard deviation sd. The syntax is $p = dnorm(x, m, sd)$.

Plotting the 500 college students example:

$$s1 := 15 \qquad m = 151$$

On the same graph, we show the standard deviation = 10 case, $s2 = 10$
$$x := 100, 102 .. 200$$
$$yc(x) := N \cdot dnorm(x, m, s1) \qquad yd(x) := N \cdot dnorm(x, m, s2)$$

We place markers at the 3σ points M1 and M2:

$$M1 := m + 3 \cdot s1 \qquad M2 := m + 3 \cdot s2$$

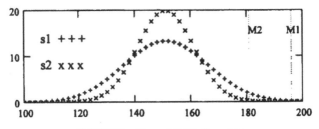

FIGURE 2. Normal distribution plots.

The RSS values of an electronic circuit are formally defined as:
(rv = random variable)

$$\text{Var}(Vo) = \sum_{i=1}^{N} \left(\frac{d}{dX_i} Vo \right)^2 \cdot \text{Var}(X_i) = \left(\sigma_{vo} \right)^2 \quad \text{Eqn (1)}$$

where Xi refers to components such as R1, C2, L3, etc. This is the variance of a function of rv's. The variance of one rv is defined as:

$$\frac{1}{N} \cdot \sum_{i} \left(x_i - x_{avg} \right)^2$$

We apply this definition to the simple voltage divider already introduced: Repeating the derivatives:

$$dmR1 := \frac{d}{dR1} G(R1, R2, E1) \quad dmR2 := \frac{d}{dR2} G(R1, R2, E1)$$

$$dmE1 := \frac{d}{dE1} G(R1, R2, E1)$$

The standard deviations for the components assumes that the components are normally distributed about the nominal values. (In some cases, this assumption may not be justified.)

$$\sigma_{R1} := \frac{R1 \cdot Tr}{3} \quad \sigma_{R2} := \frac{R2 \cdot Tr}{3} \quad \sigma_{E1} := \frac{E1 \cdot Te}{3} \quad \sigma_{R2} = 10$$

Applying the formal definition:

$$\text{var Vo} := \left(dmR1 \cdot \sigma_{R1}\right)^2 + \left(dmR2 \cdot \sigma_{R2}\right)^2 + \left(dmE1 \cdot \sigma_{E1}\right)^2$$

$$sdVo := \sqrt{\text{varVo}} \quad sdVo = 0.125$$

Then the output voltage -3σ is:

$$Vrss_1 := Vo - 3 \cdot sdVo \quad Vrss_1 = 7.124$$

The $+3\sigma$ value then is:

$$Vrss_2 := Vo + 3 \cdot sd\,Vo \quad Vrss_2 = 7.876$$

$$Vrss = \begin{bmatrix} 7.124 \\ 7.876 \end{bmatrix} \quad \Delta Vrss := Vrss - Vo \quad \Delta Vrss = \begin{bmatrix} -0.376 \\ 0.376 \end{bmatrix}$$

Comparing these numbers to the extreme values Vev from previous work:

$$Vev := \begin{bmatrix} 7.089 \\ 7.914 \end{bmatrix}$$

Note that the RSS outputs should be less than the EVA outputs.

As in EVA, we would rather have iterated expressions to save time and effort for circuits with many components. We repeat previous work to obtain the approximate sensitivities:

$$dpf := 0.0001 \quad Q := dpf \cdot identity(3) + 1$$
$$Vr_p := G(R1 \cdot Q_{p,1}, R2 \cdot Q_{p,2}, E1 \cdot Q_{p,3})$$

$$Sen_p := \left(\frac{Vr_p}{Vo} - 1\right) \cdot \frac{1}{dpf} \quad Sen^T = [-0.25 \ 0.25 \ 1]$$

$$\Delta Vr_p := Vr_p - Vo$$

$$var\ vo := \left(\frac{\Delta Vr_1}{dpf \cdot R1} \cdot \sigma_{R1}\right)^2 + \left(\frac{\Delta Vr_2}{dpf \cdot R2} \cdot \sigma_{R2}\right)^2 + \left(\frac{\Delta Vr_3}{dpf \cdot E1} \cdot \sigma_{E1}\right)^2$$

$$sdvo := \sqrt{varvo} \quad sdvo = 0.125$$

Same answer. Now we need to make some more substitutions:

$$varvo := \left(\frac{\Delta Vr_1}{dpf \cdot R1} \cdot \frac{R1 \cdot Tr}{3}\right)^2 + \left(\frac{\Delta Vr_2}{dpf \cdot R2} \cdot \frac{R2 \cdot Tr}{3}\right)^2$$
$$+ \left(\frac{\Delta Vr_3}{dpf \cdot E1} \cdot \frac{E1 \cdot Tr}{3}\right)^2$$

$$sdvo := \sqrt{varvo} \quad sdvo = 0.125$$

We see that we can cancel the component values inside the parentheses. At the same time, we will use the tolerance array T (repeated below) for the individual tolerances.

$$T = \begin{bmatrix} -0.01 & -0.01 & 0.05 \\ 0.01 & 0.01 & 0.05 \end{bmatrix}$$

$$\text{varvo} := \left(\frac{\Delta Vr_1}{dpf} \cdot \frac{T_{2,1}}{3} \right)^2 + \left(\frac{\Delta Vr_2}{dpf} \cdot \frac{T_{2,2}}{3} \right)^2 + \left(\frac{\Delta Vr_3}{dpf} \cdot \frac{T_{2,3}}{3} \right)^2$$

$$\text{sdvo} := \sqrt{\text{varvo}} \qquad \text{sdvo} = 0.125$$

We're almost there

$$k := 1 .. 2 \quad \text{(low-high counter)}$$

$$\text{varvo} := \sum_p \frac{\left(\Delta Vr_p \cdot T_{2,p} \right)^2}{\left(3 \cdot dpf \right)^2}$$

We take the square root and multiply both sides by 3 and assign this to a new variable $\Delta Vrss$.

$$\Delta Vrss_k := \frac{1}{dpf} \cdot \sqrt{\sum_p \left(\Delta Vr_p \cdot T_{k,p} \right)^2} \qquad \Delta Vrss = \begin{bmatrix} 0.376 \\ 0.376 \end{bmatrix}$$

To get the signs right, we insert $(-1)^k$:

$$\Delta Vrss_k := \frac{(-1)^k}{dpf} \cdot \sqrt{\sum_p \left(\Delta Vr_p \cdot T_{k,p} \right)^2} \qquad \qquad \text{Eqn (2)}$$

$$\Delta Vrss = \begin{bmatrix} -0.376 \\ 0.376 \end{bmatrix} \qquad Vrss := Vo + \Delta Vrss \qquad Vrss = \begin{bmatrix} 7.124 \\ 7.876 \end{bmatrix}$$

Same answers. It checks out. Substituting the sensitivities, we can get an equation that will give us a better understanding of the RSS process:

$$\Delta Vrss_k := (-1)^k \cdot Vo \cdot \sqrt{\sum_p \left(Sen_p \cdot T_{Lp} \right)^2} \qquad \text{Eqn (3)}$$

$$\Delta Vrss = \begin{bmatrix} -0.376 \\ 0.376 \end{bmatrix}$$

Hence the RSS values are strongly dependent on the *sensitivity-tolerance products*.

We can plot a Gaussian curve of the calculated RSS variations of the voltage divider and get some useful statistics:

$$N = 100 \qquad \sigma = \frac{\Delta Vrss_2}{3} \qquad v = 7, 7.02 .. 8$$

$$V = 7.5 \qquad \sigma = 0.125 \qquad g(v) = \frac{v - Vo}{\sigma}$$

$$vrss(v) := dnorm(g\,v), 0, 1) \qquad crss(v) := pnorm(g(v), 0, 1)$$

$$v3 := Vo + \Delta Vrss_2 \qquad v3 = 7.876 \qquad g(v3) = 3$$

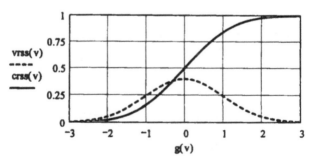

FIGURE 3. Normal distribution - yield.

What is the probability of a circuit having an output of 7.3V? Of 7.7V?

$$\text{crss}(7.3) = 0.055$$
$$\text{crss}(7.7) = 0.945$$
$$g(7.7) = 1.6$$
$$g(7.3) = -1.6$$

i.e., 5.5% of the circuits will have an output of 7.3V or less; 94.5% will have an output of 7.7V or less. If the test limits are from 7.3V to 77V, ($\sigma = 1.6$) then the *manufacturing yield* (MY) will be crss (7.7) – crss (7.3) = 0.89   or about 89%.

EVA/RSS SUMMARY (Symmetric Tolerances):

Step 1: Define circuit function and get the nominal output

$$G(R1, R2, E1) := \frac{E1 \cdot R2}{R1 + R2}$$

$$Vo := G(R1, R2, E1) \qquad Vo = 7.5$$

Step 2: Create component counter p, the Q matrix, and get sensitivities

$$p := 1..3 \qquad Q := dpf \cdot \text{identity}(3) + 1$$

$$Vrp := G(R1 \cdot Q_{p,1}, R2 \cdot Qp, 2, E1 \cdot Qp, 3$$

$$Sen_p := \left(\frac{Vr_p}{Vo} - 1 \right) \cdot \frac{1}{dpf}$$

Step 3. Create the tolerance array:

$$T := \begin{bmatrix} -Tr & -Tr & -Te \\ Tr & Tr & Te \end{bmatrix}$$

Step 4. Create the swapped tolerance array:

$$M_{k,p} := \text{if } (k = 1, \text{if } (Sen_p < 0, 1 + T_{2,p}, 1 + T_{1,p}),$$
$$\text{if } (Sen_p > 0, 1 + T_{2,p}, 1 + T_{1,p}))$$

$$M = \begin{bmatrix} 1.01 & 0.99 & 0.95 \\ 0.99 & 1.01 & 1.05 \end{bmatrix}$$

Step 5. Get the extreme values (EVA):

$$Vev_k := G(R1 \cdot M_{k,1}, R2 \cdot M_{k,2}, E1 \cdot M_{k,3}) \qquad \Delta Vev := Vev - Vo$$

$$Vev = \begin{bmatrix} 7.089 \\ 7.914 \end{bmatrix} \qquad \Delta Vev = \begin{bmatrix} -0.411 \\ 0.414 \end{bmatrix}$$

Step 6. Get RSS values using Eqn (3):

$$\Delta Vrss_k := (-1)^k \cdot Vo \cdot \sqrt{\sum_p \left(Sen_p \cdot T_{k,p}\right)^2} \qquad Vrss := Vo + \Delta Vrss$$

$$\Delta Vrss = \begin{bmatrix} -0.376 \\ 0.376 \end{bmatrix} \qquad Vrss = \begin{bmatrix} 7.124 \\ 7.876 \end{bmatrix}$$

We now perform an EVA/RSS tolerance analysis on a circuit with more components, using the summary given above as a guide.

DC DIFFERENTIAL AMPLIFIER

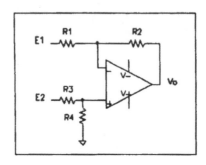

Step 1. Define circuit function and get nominal output:

$$E1 := 1 \qquad E2 := -1 \qquad R1 := 10$$
$$R2 := 100 \qquad R3 := 10 \qquad R4 := 100$$

$$G(R1, R2, R3, R4, E1, E2) := \frac{E1 \cdot \left(1 + \dfrac{R2}{R1}\right)}{1 + \dfrac{R3}{R4}} - \frac{E2 \cdot R2}{R1}$$

$$Vo := G(R1, R2, R3, R4, E1, E2) \qquad Vo = 20$$

Step 2. Create component counter p, the Q matrix, and get sensitivities:

$$Nc := 6 \qquad p := 1 .. Nc \qquad dpf := 0.0001$$

$$Q := dpf \cdot identity(Nc) + 1$$

$$Vr_p := G(R1 \cdot Q_{p,1}, R2 \cdot Q_{p,2}, R3 \cdot Q_{p,3}, \\ R4 \cdot Q_{p,4}, E1 \cdot Q_{p,5}, E2 \cdot Q_{p,6})$$

$$Vr^T = [19.9981 \quad 20.0019 \quad 19.9999$$
$$20.0001 \quad 20,001 \quad 20.001]$$

$$Sen_p := \left(\frac{Vr_p}{Vo} - 1\right) \cdot \frac{1}{dpf}$$

$$Sen^T = [-0.954 \quad 0.955 \quad -0.045 \quad 0.045 \quad 0.5 \quad 0.5]$$

Step 3. Create the tolerance array T:

$$Tr := 0.01 \qquad Te := 0.05$$

	R1	R2	R3	R4	E1	E2	(Column order same as G function order)
$T :=$	$-Tr$	$-Tr$	$-Tr$	$-Tr$	$-Te$	$-Te$	(Symmetric
	Tr	Tr	Tr	Tr	Te	Te	tolerances)

Step 4. Create the swapped tolerance array M:

$$k := 1..2$$

$$M_{k,p} := if(k = 1, if(Sen_p < 0, 1 + T_{2,p}, 1 + T_{1,p}),$$
$$if(Sen_p < 0, 1 + T_{1,p}, 1 + T_{2,p}))$$

$$M = \begin{bmatrix} 1.01 & 0.99 & 1.01 & 0.99 & 0.95 & 0.95 \\ 0.99 & 1.01 & 0.99 & 1.01 & 1.05 & 1.05 \end{bmatrix}$$

Step 5. Get the extreme values (EVA):

$$Vev_k := G(R1 \cdot M_{k,1}, R2 \cdot M_{k,2}, R3 \cdot M_{k,3},$$
$$R4 \cdot M_{k,4}, E1 \cdot M_{k,5}, E2 \cdot M_{k,6})$$

$$Vev = \begin{bmatrix} 18.624 \\ 21.424 \end{bmatrix}$$

$$\Delta Vev := Vev - Vo \qquad \Delta Vev = \begin{bmatrix} -1.376 \\ 1.424 \end{bmatrix}$$

Step 6. Get the RSS values:

$$Vrss_k := Vo \cdot \left[1 + (-1)^k \cdot \sqrt{\sum_p \left(Sen_p \cdot T_{k,p} \right)^2} \right] \qquad Vrss = \begin{bmatrix} 19.243 \\ 20.757 \end{bmatrix}$$

$$\Delta Vrss := Vrss - Vo \qquad \Delta Vrss = \begin{bmatrix} -0.757 \\ 0.757 \end{bmatrix}$$

This is how *NOT* to do the EVA:

$$evh := \frac{E1 \cdot \left(1 + T_{2,5}\right) \cdot \left[1 + \dfrac{R2 \cdot \left(1 + T_{2,2}\right)}{R1 \cdot \left(1 + T_{1,1}\right)} \right]}{1 + \dfrac{R3 \cdot \left(1 + T_{1,3}\right)}{R4 \cdot \left(1 + T_{2,4}\right)}}$$

$$- \frac{E2 \cdot \left(1 + T_{2,6}\right) \cdot R2 \cdot \left(1 + T_{2,2}\right)}{R1 \cdot \left(1 + T_{1,1}\right)}$$

$$evh = 21.424$$

ASYMMETRIC TOLERANCES

$$Vrss_k := Vo \cdot \left[1 + (-1)^k \cdot \sqrt{\sum_p \left[Sen_p \cdot \left(M_{k,p} - 1 \right) \right]^2} \right] \qquad \text{Eqn (4)}$$

$$Vrss = \begin{bmatrix} 19.243 \\ 20.757 \end{bmatrix}$$

Both Eqn (3) and Eqn (4) give the same answers with *symmetric* tolerances:

$$T = \begin{bmatrix} -0.01 & -0.01 & -0.01 & -0.01 & -0.05 & -0.05 \\ 0.01 & 0.01 & 0.01 & 0.01 & 0.05 & 0.05 \end{bmatrix}$$

$$M - 1 = \begin{bmatrix} 0.01 & -0.01 & 0.01 & -0.01 & -0.05 & -0.05 \\ -0.01 & 0.01 & -0.01 & 0.01 & 0.05 & 0.05 \end{bmatrix}$$

Sensitivity signs have no effect due to squaring, which implies that all Sen > 0, which is incorrect.

Asymmetric tolerances can occur when performing design verification analysis to show that a circuit design will meet specification over temperature and life. Environmental temperature specs are often asymmetric such as $-50°C$ to $+60°C$, for temp deltas of $60-25 = +35°C$ and $-50 - 25 = -75°C$. Then when adding the component tempco's of 50 ppm/C $\times$ ΔT for example, the overall component tolerances become asymmetric.

We simulate this by increasing the high tolerances of R1 and R3 from 1% to 3%. The low tolerance remains at -1%. Note that R1 and R3 have negative sensitivities.

$$T := \begin{bmatrix} -Tr & -Tr & -Tr & -Tr & -Te & -Te \\ 0.03 & Tr & 0.03 & Tr & Te & Te \end{bmatrix}$$

$$T = \begin{bmatrix} -0.01 & -0.01 & -0.01 & -0.01 & -0.05 & -0.05 \\ 0.03 & 0.01 & 0.03 & 0.01 & 0.05 & 0.05 \end{bmatrix}$$

Then using Eqn (3) and naming it Vrs instead of Vrss:

$$Vrs_k := Vo \cdot \left[1 + (-1)^k \cdot \sqrt{\sum_p \left(Sen_p \cdot T_{k,p} \right)^2} \right]$$

$$Vrs = \begin{bmatrix} 19.243 \\ 20.93 \end{bmatrix} \qquad \Delta Vrs := Vrs - Vo \qquad \Delta Vrs = \begin{bmatrix} -0.757 \\ 0.93 \end{bmatrix}$$

Note that the $+\Delta$ is higher in absolute magnitude than is the $-\Delta$.
Swap new tolerances and use Eqn (4):

$$M_{k,p} := if\,(k = 1, if\,(Sen_p < 0, 1 + T_{2,p}, 1 + T_{1,p}),$$
$$if\,(Sen_p < 0, 1 + T_{1,p}, 1 + T_{2,p}))$$

$$M - 1 = \begin{bmatrix} 0.03 & -0.01 & 0.03 & -0.01 & -0.05 & -0.05 \\ -0.01 & 0.01 & -0.01 & 0.01 & 0.05 & 0.05 \end{bmatrix}$$

$$Vrss_k := Vo \cdot \left[1 + (-1)^k \cdot \sqrt{\sum_p \left[Sen_p \cdot \left(M_{k,p} - 1 \right) \right]^2} \right]$$

$$Vrss = \begin{bmatrix} 19.07 \\ 20.757 \end{bmatrix} \qquad \Delta Vrss := Vrss - Vo$$

$$\Delta Vrss = \begin{bmatrix} -0.93 \\ 0.757 \end{bmatrix} \quad \text{Now the } +\Delta \text{ is smaller than the } -\Delta$$

To get a benchmark, the EVA with the asymmetric tolerances is computed:

$$\text{Vev}_k := G(R1 \cdot M_{k,1}, R2 \cdot M_{k,2}, R3 \cdot M_{k,3},$$
$$R4 \cdot M_{k,4}, E1 \cdot M_{k,5}, E2 \cdot M_{k,6})$$

$$\text{Vev} = \begin{bmatrix} 18.262 \\ 21.424 \end{bmatrix} \qquad \Delta\text{Vev} := \text{Vev} - \text{Vo} \qquad \Delta\text{Vev} = \begin{bmatrix} -1.738 \\ 1.424 \end{bmatrix}$$

$$\Delta\text{Vrss} = \begin{bmatrix} -.093 \\ 0.757 \end{bmatrix} \qquad \Delta\text{Vrs} = \begin{bmatrix} -0.757 \\ 0.93 \end{bmatrix}$$

The $-\Delta\text{Vev}$ is pulled down farther that the $+\Delta\text{Vev}$ due to the increased tolerances of R1 and R3 of 3%. (Both have negative sensitivity signs.) The same thing happens with ΔVrss using Eqn (4), but not with ΔVrs using Eqn (3).

We now compare ratios to show that ΔVrss is a better estimate than ΔVrs.

Define ratio function:

$$r(a, b) := \frac{a - b}{a - \text{Vo}}$$

$$r(\text{Vev}_1, \text{Vrss}_1) = 0.465 \qquad r(\text{Vev}_2, \text{Vrss}_2) = 0.468$$

Ratios approximately equal when using Eqn (4)

$$r(\text{Vev}_1, \text{Vrs}_1) = 0.564 \qquad r(\text{Vev}_2, \text{Vrs}_2) = 0.347$$

Ratios not equal when using Eqn (3)

This will be verified when MCA methods are covered. (Get in the habit of using Eqn (4) even though the tolerances may be symmetric. It covers both.)

To recap:

Step 1: Define circuit G-function and get the nominal output

$$G(R1, R2, R3, R4, E1, E2) := \frac{E1 \cdot \left(1 + \dfrac{R2}{R1}\right)}{1 + \dfrac{R3}{R4}} - \frac{E2 \cdot R2}{R1}$$

$$Vo := G(R1, R2, R3, R4, E1, E2)$$

Step 2: Create component counter p, the Q matrix, and get sensitivities

$$Nc := 6 \quad p := 1..Nc \quad dpf := 0.0001 \quad Q := dpf \cdot identity(Nc) + 1$$

$$Vr_p := G(R1 \cdot Q_{p,1}, R2 \cdot Q_{p,2}, R3 \cdot Q_{p,3}, R4 \cdot Q_{p,4}, Eq \cdot Q_{p,5}, E2 \cdot Q_{p,6})$$

$$Sen_p := \left(\frac{Vr_p}{Vo} - 1\right) \cdot \frac{1}{dpf}$$

Step 3: Create tolerance array T

$$Tr := 001 \quad Te := 0.05 \quad T := \begin{bmatrix} -Tr & -Tr & -Tr & -Tr & -Te & -Te \\ Tr & Tr & Tr & Tr & Te & Te \end{bmatrix}$$

Step 4: Create the swapped tolerance array:

$$k := 1..2$$

$$M_{k,p} := if(k = 1, \; if(Sen_p < 0, 1 + T_{2,p}, 1 + T_{1,p}),$$
$$if(Sen_p < 0, 1 + T_{1,p}, 1 + T_{2,p}))$$

Step 5: Get extreme values (EVA)

$$Vev_k := G(R1 \cdot M_{k,1}, R2 \cdot M_{k,2}, R3 \cdot M_{k,3}, R4 \cdot M_{k,4}, E1 \cdot M_{k,5}, E2 \cdot M_{k,6})$$

$$Vev = \begin{bmatrix} 18.624 \\ 21.424 \end{bmatrix}$$

Step 6: Get RSS values

$$Vrss_k := Vo \cdot \left[1 + (-1)^k \cdot \sqrt{\sum_p \left[Sen_p \cdot \left(M_{k,p} - 1 \right) \right]^2} \right] \quad Vrss = \begin{bmatrix} 19.243 \\ 20.757 \end{bmatrix}$$

When specifying test limits, don't forget the tolerance of the measuring device (DVM, scope, etc.). Assume a DVM with a dc tolerance of 2%:

$$Tm := 0.02$$

$$Vtev_k := [1 + (-1)^k \cdot Tm] \cdot Vev_k \quad Vtrss_k := [1 + (-1)^k \cdot Tm] \cdot Vrss_k$$

New limits including test equipment tolerances:

$$Vtev = \begin{bmatrix} 18.251 \\ 21.853 \end{bmatrix} \quad Vtrss = \begin{bmatrix} 18.858 \\ 21.172 \end{bmatrix}$$

GENERAL RSS/EVA EQUATION SEQUENCE

$$Vo = G(X1, X2, X3, \ldots, X_{Nc})$$

$$p = 1 .. Nc \quad \text{(same for every circuit)}$$

$$dpf = 0.0001 \quad \text{(same)}$$

$$Q = dpf \cdot identity\,(Nc) + 1 \quad \text{(same)}$$

$$Vr_p = G\,(X1 \cdot Q_{p,1}, X2 \cdot Q_{p,2}, X3 \cdot Q_{p,3}, \ldots, X_{Nc} \cdot Q_{p,Nc})$$

$$Sen_p = \left(\frac{Vr_p}{Vo} - 1\right) \cdot \frac{1}{dpf} \quad \text{(same)}$$

$$T = \begin{bmatrix} -T1 & -T2 & -T3 & -T_{Nc} \\ T1 & T2 & T3 & T_{Nc} \end{bmatrix}$$

$$k = 1, 2 \quad \text{(same)}$$

$$M_{k,p} = if\,(k = 1, if\,(Sen_p < 0, 1 + T_{2,p}, 1 + T_{1,p}),$$
$$if\,(Sen_p \geq 0, 1 + T_{2,p}, 1 + T_{1,p})) \quad \text{(same)}$$

$$Vev_k = G\,(X1 \cdot M_{k,1}, X2 \cdot M_{k,2}, X3 \cdot M_{k,3}, \ldots, X_{Nc} \cdot M_{k,Nc})$$

$$Vrss_k := Vo \cdot \left[1 + (-1)^k \cdot \sqrt{\sum_p \left[Sen_p \cdot \left(M_{k,p} - 1\right)\right]^2}\,\right] \quad \text{(same)} \qquad \text{Eqn (4)}$$

COMPARING MATHCAD EVA WITH Spice.WCASE Analysis

We now perform a "reality check" to see how close the Mathcad routines come to a Spice analysis of the same circuit. (Note: there does not appear to be any way to tolerance the input voltages using Spice's .MODEL statements. Hence only four resistor components are toleranced below to compare with the Spice answers.)

$$E1 := 1 \quad E2 := -1 \quad R1 := 10 \quad R2 := 100 \quad R3 := 10 \quad R4 := 100$$

$$G(R1, R2, R3, R4, E1, E2) := \frac{E1 \cdot \left(1 + \frac{R2}{R1}\right)}{1 + \frac{R3}{R4}} - \frac{E2 \cdot R2}{R1}$$

$$Vo := G(R1, R2, R3, R4, E1, E2) \qquad Vo = 20$$

(Note that dpf is increased to 0.001 to get the same answers that Spice does).

$$Nc := 4 \qquad p := 1 .. Nc \qquad dpf := 0.001 \qquad Q := dpf \cdot identity(Nc) + 1$$

$$Vr_p := G(R1 \cdot Q_{p,1}, R2 \cdot Q_{p,2}, R3 \cdot Q_{p,3}, R4 \cdot Q_{p,4}, E1, E2)$$

$$Vr^T - Vo = [-0.0191 \quad 0.0191 \quad -9.0901 \cdot 10^{-4} \quad 9.0827 \cdot 10^{-4}]$$

Compare with Spice output on next page.

$$Sen_p := \left(\frac{Vr_p}{Vo} - 1\right) \cdot \frac{1}{dpf} \qquad Sen^T = [-0.954 \quad 0.955 \quad -0.045 \quad 0.045]$$

$$Tr := 0.02 \qquad T := \begin{bmatrix} -Tr & -Tr & -Tr & -Tr \\ Tr & Tr & Tr & Tr \end{bmatrix} \qquad k := 1 .. 2$$

$$M_{k,p} := \text{if}(k = 1, \text{if}(Sen_p > 0, 1 + T_{1,p}, 1 + T_{2,p}), \\ \text{if}(Sen_p > 0, 1 + T_{2,p}, 1 + T_{1,p}))$$

$$Vev_k := G(R1 \cdot M_{k,1}, R2 \cdot M_{k,2}, R3 \cdot M_{k,3}, R4 \cdot M_{k,4}, E1,$$

$$Vev = \begin{bmatrix} 19.216 \\ 20.816 \end{bmatrix}$$

$$\frac{Vev_2}{Vo} = 104.08 \cdot \%$$

The following is extracted from the spice *.out text file. Note the percent change at the bottom of the page is also 104.08%. Also shown on page 32 is a sensitivity (.SENS) run of the same circuit. This shows that the sensitivities compare favorably.

```
***************** 01/23/99 07:41:25 ***************
******** NT Evaluation PSpice (July 1997) ********

  EVA OF DIFF AMP
  ****        CIRCUIT DESCRIPTION
V2 1 0 DC -1
V1 4 0 DC 1
R1 1 2 RA 10
R2 2 3 RA 100
R3 4 5 RA 10
R4 5 0 RA 100
E1 3 0 5 2 1E6
*.STEP V2 LIST -1 -1.05
.MODEL RA RES(R=1 DEV/GAUSS 0.667%)
* 3sigma = 2%
.WCASE DC V(3) YMAX VARY DEV
.DC V1 LIST 1
.OPTIONS NOMOD NOECHO NOPAGE


  ****        SORTED DEVIATIONS OF V(3)
              TEMPERATURE = 27,000 DEG C
              SENSITIVITY SUMMARY


Mean Deviation =    3.8147E-06
Sigma          =       .0135
```

Compare the following with Vr in Mathcad

```
RUN    MAX DEVIATION FROM NOMINAL
R2 RA R       .0191  (1.41 sigma) higher at V1 = 1
              (    .9545% change per 1% change in
                   Model Parameter)


R1 RA R       .0191  (1.41 sigma) lower  at V1 = 1
              (    .9537% change per 1% change in
                   Model Parameter)


R3 RA R    909.8100E-06  ( .07 sigma) lower at V1 = 1
              (    .0455% change per 1% change in
                   Model Parameter)


R4 RA R    907.9000E-06  ( .07 sigma) higher at V1 = 1
              (    .0454% change per 1% change in
                   Model Parameter)


              WORST CASE ALL DEVICES


DEVICE          MODEL          PARAMETER        NEW VALUE
R1              RA             R                     .98
R2              RA             R                    1.02
R3              RA             R                     .98
R4              RA             R                    1.02


              WORST CASE ALL SUMMARY


RUN             MAX DEVIATION FROM NOMINAL

ALL DEVICES    .8167 higher at V1 = 1
               (104.08% of Nominal)
```

```
**************** 11/24/98 20:37:07 ***************

******** NT Evaluation PSpice (July 1997) ********

SENS OF DIFF AMP

V2 1 0 DC -1
V1 4 0 DC 1
R1 1 2 10
R2 2 3 100
R3 4 5 10
R4 5 0 100
E1 3 0 5 2 1E6
 .SENS V(3)
 .OPTIONS NOMOD NOECHO NOPAGE
```

```
  NODE          VOLTAGE     NODE          VOLTAGE

(    1)        -1.0000    (    2)          .9091
(    3)        20.0000    (    4)         1.0000
(    5)          .9091
```

DC SENSITIVITIES OF OUTPUT V(3)

ELEMENT NAME	ELEMENT VALUE	ELEMENT SENSITIVITY (VOLTS/UNIT)	NORMALIZED SENSITIVITY (VOLTS/PERCENT)
R1	1.000E+01	-1.909E+00	-1.909E-01
R2	1.000E+02	1.909E-01	1.909E-01
R3	1.000E+01	-9.091E+02	-9.091E-03
R4	1.000E+02	9.091E-03	9.091E-03
V2	-1.000E+00	-1.000E+01	1.000E-01
V1	1.000E+00	1.000E+01	1.000E-01

Normalize R1 sensitivities to compare with Mathcad:

$$-1.909 \, \frac{V}{\Omega} \cdot \frac{10\Omega}{20V} = -0.954$$

$$-0.1909 \, \frac{V}{\%} \cdot \frac{100\%}{20V} = -0.954$$

COMPARING CIRCUIT ANALYSIS METHODS

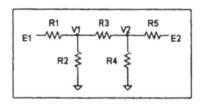

E1 := 10	E2 := 20	R1 := 20	R2 := 100
R3 := 10	R4 := 100	R5 := 50	

Two new sets of input voltages:

E1a := 15   E2a := −15   E1b := 8   E2b := −7.5

Transfer function method:

$$V2 := \frac{\left(\left(E1 \cdot R5 + E2 \cdot (R1+R3)\right) \cdot R2 + E2 \cdot R1 \cdot R3\right) \cdot R4}{\left((R1+R3+R4) \cdot R5 + R4 \cdot (R1+R3)\right)} \\ \cdot R2 + R1 \cdot \left(R5 \cdot (R3+R4) + R3 \cdot R4\right)$$

$$V2 = 10.556$$

A considerable amount of algebra is required to obtain this transfer function — for only one node voltage!

Matrix method using Mathcad programming:

$$G(R1, R2, R3, R4, R5, E1, E2) := \begin{vmatrix} A \leftarrow \begin{bmatrix} \dfrac{1}{R1} + \dfrac{1}{R2} + \dfrac{1}{R3} & \dfrac{-1}{R3} \\[2ex] \dfrac{-1}{R3} & \dfrac{1}{R3} + \dfrac{1}{R4} + \dfrac{1}{R5} \end{bmatrix} \\[6ex] B \leftarrow \begin{bmatrix} \dfrac{E1}{R1} & \dfrac{E1a}{R1} & \dfrac{E1b}{R1} \\[2ex] \dfrac{E2}{R5} & \dfrac{E2a}{R5} & \dfrac{E2b}{R5} \end{bmatrix} \\[4ex] C \leftarrow A^{-1} \cdot B \\[1ex] C \end{vmatrix}$$

$$C := G(R1, R2, R3, R4, R5, E1, E2) \qquad C = \begin{bmatrix} 9.722 & 6.25 & 3.426 \\ 10.556 & 2.5 & 1.481 \end{bmatrix}$$

The matrix method gives all node voltages for multiple inputs. In general, the matrix method is preferred. However, when available, the transfer function may provide insight into circuit operation.

Note that element $A_{1,1}$ is the sum of all the conductances connected to node V1. Correspondingly, element $A_{2,2}$ is the sum of all conductances connected to node V2. For $A_{1,2}$ and $A_{2,1}$, R3 is the resistor connected *between* nodes V1 and V2. This procedure is discussed further on the next page.

MORE COMPLICATED CIRCUITS

The worst-case analysis steps are straightforward given that a transfer function (G-function) of circuit can be created. If it is impractical to obtain due to the large order of the circuit, then the matrix method must be used.

As alluded to on the previous page, there is a circuit analysis method that is given in just about every under-graduate text on circuit or network analysis. It is a mnemonic method used for passive circuits derived from nodal analysis. No controlled sources could be used as this would destroy the symmetry of the coefficient matrix and the mnemonic method.

The A coefficient matrix is created by inspection of the circuit. On the *main diagonal* of the matrix, all the *reciprocal impedances are added* together if they are connected to that particular node represented by the row = column of the matrix. All *off-diagonal elements* are the *negative reciprocal impedances* (admittances) *connected between the nodes*. As will be demonstrated, this method can be used with controlled sources such as opamps and discrete transistors. The only requirement is that the analysis be linear.

Following is an example of a circuit the author was required to worst-case analyze while working for a company on the International Space Station project. Management wanted an RSS analysis since they believed EVA went out with the Cold War. In addition, the temperature extremes of the space station environment, +35°C to −80°C, created asymmetric tolerances for the components.

The circuit is an RTD (resistive-temperature detector) temperature sensing circuit with an output of −5V when the RTD resistor was 1K in value (0°C), and +5V when the RTD was 2K (+260°C). Accuracy specifications were ≤ ± 5°C. The RTD output is approximately linear and is given by:

$$RT = Ro\left(\frac{T}{260} + 1\right), \quad \text{where } Ro = 1K$$

No transfer function is available due to the large number of unknown nodes. Normally, Spice would come to the rescue, but Spice was ruled out on two counts: (1) the task requires RSS analysis and Spice does not currently have that capability, and (2) the component tolerances are asymmetric and the Spice .WCASE feature allows only symmetric tolerances. Hence the matrix method can be used to analyze this circuit.

The RTD circuit schematic is shown below:

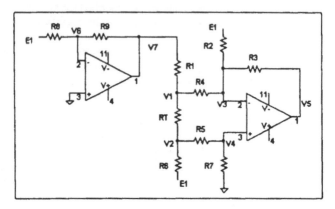

The RTD A matrix created directly from the above schematic is shown below.

$$
A = \begin{bmatrix}
\dfrac{1}{R1} + \dfrac{1}{R4} + \dfrac{1}{RT} & \dfrac{-1}{RT} & \dfrac{-1}{R4} & 0 & 0 & 0 & \dfrac{-1}{R1} \\[2mm]
\dfrac{-1}{RT} & \dfrac{1}{R5} + \dfrac{1}{R6} + \dfrac{1}{RT} & 0 & \dfrac{-1}{R5} & 0 & 0 & 0 \\[2mm]
\dfrac{-1}{R4} & 0 & \dfrac{1}{R2} + \dfrac{1}{R3} + \dfrac{1}{R4} & 0 & \dfrac{-1}{R3} & 0 & 0 \\[2mm]
0 & \dfrac{-1}{R5} & 0 & \dfrac{1}{R5} + \dfrac{1}{R7} & 0 & 0 & 0 \\[2mm]
0 & 0 & Ao & -Ao & 1 & 0 & 0 \\[2mm]
0 & 0 & 0 & 0 & 0 & \dfrac{1}{R8} + \dfrac{1}{R9} & \dfrac{-1}{R9} \\[2mm]
0 & 0 & 0 & 0 & 0 & Ao & 1
\end{bmatrix}
$$

For example, to show how row 2 of the A matrix is created, the KCL equation for node V2 is:

$$\frac{E1 - V2}{R6} = \frac{V2 - V4}{R5} + \frac{V2 - V1}{RT}$$

Placing all the unknowns on the left-hand side and the known inputs on the right-hand side and factoring gives:

$$V2 \cdot \left(\frac{1}{R5} + \frac{1}{R6} + \frac{1}{RT}\right) - \frac{V1}{RT} - \frac{V4}{R5} = \frac{E1}{R6}$$

Hence, for the 2nd row (the second equation) columns 1, 2, and 4, for nodes V1, V2, and V4 respectively, have the same coefficients as the above equation.

For the B input column vector, one only has to assign the partial currents caused by the independent source E1. Hence $\frac{E1}{R6}$ is replaced in row 2 of the B vector: This is transformed to a row vector to save print space.

$$B = \left[0 \; \frac{E1}{R6} \; \frac{E1}{R2} \; 0 \; 0 \; \frac{E1}{R8} \; 0\right]^{T}$$

For row 5, the opamp is a voltage-controlled voltage source (VCVS) and is given by the simple first order dc equation as, using open loop gain Ao = 10^6 V/V:

$$V5 = Ao \cdot (V4 - V3)$$

Placing all unknowns of the left-hand side: V5 – Ao·V4 + Ao·V3 = 0
Hence the coefficient in column 5 is 1, that in column 4 is –Ao, and in column 3, +Ao. The zero on the right-hand side is placed in row 5 of the B vector. Hence this method of circuit analysis is simple book-keeping — placing numbers in rows and columns per the schematic.

The RSS method described up to this point has been with the assumption of Gaussian distribution of component tolerances. Introduced on page 41 is a method for determing the output spread assumming a uniform distribution of component tolerances. See the Appendix for a derivation of this method.

The EVA/RSS analysis of this circuit follows:

MATHCAD EVA/RSS ANALYSIS OF AN RTD CIRCUIT

Define circuit G-function and get nominal output. All resistors are KΩ's:

$$
\begin{array}{lll}
R2 := 34.8 & R4 := 9.09 & E1 := 5 \\
5R6 := 4.53 & RT := 1.915 & R9 := 20 \\
& & \\
R1 := 4.53 & R3 := 132 & Ao := 10^6 \\
R5 := 9.09 & R7 := 27.4 & R8 := 20
\end{array}
$$

$G(R1, R2, R3, R4, R5, R6, R7, R8, R9, RT, E1) :=$

$$
\begin{vmatrix}
A \leftarrow \begin{bmatrix}
\frac{1}{R1}+\frac{1}{R4}+\frac{1}{RT} & \frac{-1}{RT} & \frac{-1}{R4} & 0 & 0 & 0 & \frac{-1}{R1} \\
\frac{-1}{RT} & \frac{1}{R5}+\frac{1}{R6}+\frac{1}{RT} & 0 & \frac{-1}{R5} & 0 & 0 & 0 \\
\frac{-1}{R4} & 0 & \frac{1}{R2}+\frac{1}{R3}+\frac{1}{R4} & 0 & \frac{-1}{R3} & 0 & 0 \\
0 & \frac{-1}{R5} & 0 & \frac{1}{R5}+\frac{1}{R7} & 0 & 0 & 0 \\
0 & 0 & Ao & -Ao & 1 & 0 & 0 \\
0 & 0 & 0 & 0 & 0 & \frac{1}{R8}+\frac{1}{R9} & \frac{-1}{R9} \\
0 & 0 & 0 & 0 & 0 & Ao & 1
\end{bmatrix} \\
B \leftarrow \begin{bmatrix} 0 & \frac{E1}{R6} & \frac{E1}{R2} & 0 & 0 & \frac{E1}{R8} & 0 \end{bmatrix}^T \\
C \leftarrow \text{lsolve}(A, B) \\
Vo \leftarrow C_5 \\
Vo
\end{vmatrix}
$$

$Vo := G(R1, R2, R3, R4, R5, R6, R7, R8, R9, RT, E1)$ $Vo = 4.326$

Create component counter p, the Q matrix, and get sensitivities:

$Nc := 11$ $p := 1..Nc$ $dpf := 0.0001$ $Q := dpf \cdot identity(Nc) + 1$

$$V5p := G(R1 \cdot Q_{p,1},\ R2 \cdot Q_{p,2},\ R3 \cdot Q_{p,3},\ R4 \cdot Q_{p,4},$$
$$R5 \cdot Q_{p,5},\ R6 \cdot Q_{p,6},\ R7 \cdot Q_{p,7},\ R8 \cdot Q_{p,8},$$
$$R9 \cdot Q_{p,9},\ RT \cdot Q_{p,10},\ E1 \cdot Q_{p,11})$$

$$Sen_p := \left(\frac{V5_p}{Vo} - 1\right) \cdot \frac{1}{dpf}$$

$$Sen^T = [\begin{matrix} -2.412 & 3.679 & 0.814 & -4.133 & -0.822 & -2.109 \\ 0.893 & -2.701 & 2.701 & 4.089 & 1] \end{matrix}$$

Create tolerance array T:

$Tinit := 0.001$ $Tlife := 0.002$ $ppm := 10^{-6}$ $TC1 := 50 \cdot ppm$
$TC2 := 25 \cdot ppm$ $Thi := Tinit + Tlife + 35 \cdot TC1$
$Tlo := -Tinit - Tlife - 80 \cdot TC1$

$Trhi := 8.1 \cdot 10^{-4}$ $Trlo := Trhi$ $Treflo := -0.02 - 80 \cdot TC2$
$Trefhi := 0.02 + 35 \cdot TC2$

$$T := \begin{bmatrix} Tlo & Tlo & Tlo & Tlo & Tlo & Tlo & Tlo & Tlo & Tlo & Trlo & T \\ Thi & Thi & Thi & Thi & Thi & Thi & Thi & Trhi & Thi & Trhi & T \end{bmatrix}$$

$$T = \begin{bmatrix} -0.7 & -0.7 & -0.7 & -0.7 & -0.7 & -0.7 & -0.7 & -0.7 & -0.7 & -0.081 & -2.2 \\ 0.475 & 0.475 & 0.475 & 0.475 & 0.475 & 0.475 & 0.475 & 0.081 & 0.475 & 0.081 & 2.087 \end{bmatrix} \cdot \%$$

Note asymmetric tolerances:

Create the swapped tolerance array M:

$$k := 1 .. 2$$

$$M_{k,p} := if(k = 1, if(Sen_p > 0, 1 + T_{1,p}, 1 + T_{2,p}),$$
$$if(Sen_p > 0, 1 + T_{2,p}, 1 + T_{1,p}))$$

$$M = \begin{bmatrix} 1.005 & 0.993 & 0.993 & 1.005 & 1.005 & 1.005 & 0.993 & 1.001 & 0.993 & 0.999 & 0.978 \\ 0.993 & 1.005 & 1.005 & 0.993 & 0.993 & 0.993 & 1.005 & 0.993 & 1.005 & 1.001 & 1.021 \end{bmatrix}$$

Get the extreme values (EVA):

$$Vev_k := G(R1 \cdot M_{k,1}, R2 \cdot M_{k,2}, R3 \cdot M_{k,3}, R4 \cdot M_{k,4},$$
$$R5 \cdot M_{k,5}, R6 \cdot M_{k,6}, R7 \cdot M_{k,7}, R8 \cdot M_{k,8},$$
$$R9 \cdot M_{k,9}, RT \cdot M_{k,10}, E1 \cdot M_{k,11})$$

$$Vo = 4.326 \quad Vev = \begin{bmatrix} 3.782 \\ 4.985 \end{bmatrix} \quad \Delta Vev := Vev - Vo \quad Vev = \begin{bmatrix} -0.544 \\ 0.659 \end{bmatrix}$$

Get the RSS values for both Gaussian input and uniform input.

Gaussian input:

$$Vrss_k := Vo \cdot \left[1 + (-1)^k \cdot \sqrt{\sum_p \left[Sen_p \cdot \left(M_{k,p} - 1 \right) \right]^2} \right]$$

$$\Delta Vrss := Vrss - Vo \quad Vrss = \begin{bmatrix} 4.122 \\ 4.55 \end{bmatrix} \quad \Delta Vrss = \begin{bmatrix} -0.204 \\ 0.224 \end{bmatrix}$$

Uniform input:

$$Vrsu := Vo + \sqrt{3} \cdot \Delta Vrss$$

$$\Delta Vrss := Vrsu - Vo \qquad Vrsu = \begin{bmatrix} 3.972 \\ 4.714 \end{bmatrix} \qquad \Delta Vrsu = \begin{bmatrix} -0.353 \\ 0.388 \end{bmatrix}$$

Optional: summarize in one output array

$$Va := augment(Vev, Vrsu) \qquad Vb := augment(Va, Vrss)$$

$$Vev \quad Vrsu \quad Vrss$$

$$Vb = \begin{bmatrix} 3.782 & 3.972 & 4.122 \\ 4.985 & 4.714 & 4.55 \end{bmatrix} \qquad Vo = 4.326$$

$$\Delta Vb := Vb - Vo \qquad \Delta Vb = \begin{bmatrix} -0.544 & -0.353 & -0.204 \\ 0.659 & 0.388 & 0.224 \end{bmatrix}$$

Since the specified accuracy is +/–5°C, a 10 degree spread, and the (lowest) Vrss spread is 0.224 + 0.204 = 0.448V, the accuracy is: Temperature-voltage gradient times:

$$Vrss = \frac{260}{10} \cdot 0.224 = 5.82°C$$

The circuit fails to meet the specification.

TOLERANCE ANALYSIS OF AC CIRCUITS

Prior to any tolerance analysis of ac circuits, a simple circuit will be analyzed in Mathcad to illustrate the general method. The circuit used is a bandwidth limited differentiator. In this circuit, the inverse of a series RC circuit (admittance) is developed as follows:

$$Z(s) = R + \frac{1}{sC} = \frac{sRC+1}{sC}, \quad \frac{1}{Z(s)} = Y(s) = \frac{sC}{sRC+1}$$

The nodal equations are:

$$\frac{E1 - Vn}{Z(s)} = \frac{Vn - Vo}{R2}$$

or

$$Vn\left(\frac{1}{Z(s)} + \frac{1}{R2}\right) - \frac{Vo}{R2} = \frac{E1}{Z(s)}$$

and

$$Vo = -Aol \cdot Vn$$

$$Aol \cdot Vn + Vo = 0$$

where Aol is the opamp open-loop gain $\approx 10^6$ V/V. The Mathcad analysis follows.

MATHCAD AC ANALYSIS METHOD

BANDWIDTH LIMITED DIFFERENTIATOR

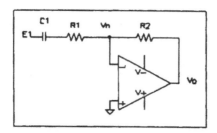

$$K := 10^3 \quad R1 := 1 \cdot K \quad R2 := 10 \cdot K$$

$$nF := 10^{-9} \quad C1 := 10 \cdot nF$$

$$db(x) := 20 \cdot \log(|x|)$$

$$BF := 3 \quad BF = \text{beginning frequency}$$

$$ND := 4 \quad ND = \text{number of decades}$$

$$PD := 20 \quad PD = \text{points per decade}$$

$$i := 1 .. ND \cdot PD + 1 \quad \text{(Frequency counter)}$$

$$L_i := BF + \frac{i-1}{PD} \quad \text{(Log frequency increment)}$$

$$F_i := 10^{L_i} \quad j := \sqrt{-1} \quad \text{(MathCad complex frequency)}$$

$$s_. := 2 \cdot \pi \cdot F_i \cdot j \quad (s = j\omega)$$

$$\omega1 := 2 \cdot \pi \cdot 100 \quad \omega2 := 2 \cdot \pi \cdot 2 \cdot 10^6 \quad (\omega1, \omega2 \text{ are opamp poles})$$

$$Aol_i := \frac{10^5}{\left(1 + \frac{s_i}{\omega1}\right) \cdot \left(1 + \frac{s_i}{\omega2}\right)} \quad \text{(Opamp open loop gain)}$$

$$A_i := \begin{bmatrix} \dfrac{s_i \cdot C1}{s_i \cdot R1 \cdot C1 + 1} + \dfrac{1}{R2} & -\dfrac{1}{R2} \\ Aol_i & 1 \end{bmatrix} \quad B_i := \begin{bmatrix} \dfrac{s_i \cdot C1}{s_i \cdot R1 \cdot C1 + 1} \\ 0 \end{bmatrix}$$

Matrices setup per mnemonic method.

Use Mathcad's lsolve routine:

$$C_i := lsolve(A_i, B_i) \quad Vo_i := db\left[(C_i)_2\right] \quad Al_i := db(Aol_i)$$

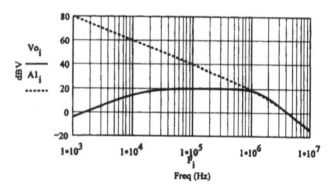

FIGURE 4. AC output plot.

TOLERANCE ANALYSIS OF A BANDPASS FILTER

With this background in Mathcad ac analysis methods, we are ready to perform a RSS/EVA analysis of a bandpass filter (BPF) shown below. This type of BPF is known as multiple feedback BPF since there are two feedback paths via R3 and C2.

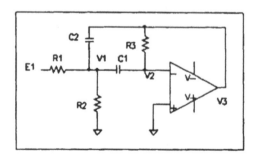

The filter has a center frequency of $f_0 = 500$ Hz and a pole Q of 20. The component values then are $R1 = 6.34K$, $R2 = 80.6\Omega$, $R3 = 127K$.

The capacitor values are C1 = C2 = 0.1 uF. The MathCad setup using a linear frequency sweep is as follows:

BF = 400 (Beginning frequency in Hz)

DF = 2 (Frequency increment in Hz)

LF = 600 (Last Frequency in Hz)

Increment counter:　　　Frequency in Hz:　　Laplace transform:

$$i = 1 \ldots \left(\frac{LF - BF}{DF} \right) + 1 \qquad F_i = BF + DF(i-1) \qquad s_i = 2\pi F_i j \quad (s = j\omega)$$

We set up the A matrix and B vector per the mnemonic method.

$$A_i = \begin{bmatrix} \dfrac{1}{R1} + \dfrac{1}{R2} + s_i C1 + s_i C2 & -s_i C1 & -s_i C2 \\[2mm] -s_i C1 & \dfrac{1}{R3} + s_i C1 & -\dfrac{1}{R3} \\[2mm] 0 & Ao & 1 \end{bmatrix}$$

$$B_i = \begin{bmatrix} \dfrac{1}{R1} & 0 & 0 \end{bmatrix}^T$$

The solution is obtained via Mathcad's lsolve function:
C = lsolve (A, B), and then Vo = C_3.

Set resistor tolerances at 2%; Tr = 0.02; and capacitor tolerances at 10%; Tc = 0.1. (Symmetric tolerances) and form T array:

$$T = \begin{bmatrix} -Tr & -Tr & -Tr & -Tc & -Tc \\ Tr & Tr & Tr & Tc & Tc \end{bmatrix}$$

After specifying the component tolerances in the T array, we continue with the process as before:

$Nc = 5 \quad p = 1...Nc \quad dpf = 0.0001 \quad Q = dpf \cdot identity(Nc) + 1$

Now we can perturbate the G-function instead of each element in the A and B arrays:

$$Vr_{i,p} = G(R1 \cdot Q_{p,1}, R2 \cdot Q_{p,2}, R3 \cdot Q_{p,3}, C1 \cdot Q_{p,4}, C2 \cdot Q_{p,5}, s)$$

The sensitivities are:

$$Sen_{i,p} = \left(\frac{Vr_{i,p}}{Vo_i} - 1 \right) \cdot \frac{1}{dpf}$$

The M array is separated into a L (for low) array and a H (for high) array as follows:

$$L_{i,p} = if(Sen_{i,p} > 0, 1 + T_{1,p}, 1 + T_{2,p})$$

$$H_{i,p} = if(Sen_{i,p} > 0, 1 + T_{2,p}, 1 + T_{1,p})$$

The RSS output is obtained with the usual equation. In the Mathcad file, an if statement is used to choose either L or H prior to insertion into the RSS equation. EVA low and high are then:

$$Vev_{1,i} = G(R1 \cdot L_{i,1}, R2 \cdot L_{i,2}, R3 \cdot L_{i,3}, C1 \cdot L_{i,4}, C2 \cdot L_{i,5})$$

$$Vev_{2,i} = G(R1 \cdot H_{i,1}, R2 \cdot H_{i,2}, R3 \cdot H_{i,3}, C1 \cdot H_{i,4}, C2 \cdot H_{i,5})$$

BANDPASS FILTER: $Q = 20, \; fo = 500 \; Hz$

$K := 10^3 \qquad uF := 10^{-6} \qquad R1 := 6.34 \cdot K \qquad R2 := 80.6$

$R3 := 127 \cdot K \qquad C1 := 0.1 \cdot uF \qquad C2 := C1 \qquad BF := 400$

$$LF := 600 \qquad DF := 2 \qquad i := 1..\left(\frac{LF - BF}{DF}\right) + 1$$

$$F_i := BF + DF \cdot (i - 1)$$

$$P := 2 \cdot \pi \qquad s_i := P \cdot F_i \cdot \sqrt{-1} \qquad Ao := 10^6$$

Begin G-function program:

$$G(R1, R2, R3, C1, C2, s) :=$$

$$
\begin{vmatrix}
A_i \leftarrow \begin{bmatrix}
\dfrac{1}{R1} + \dfrac{1}{R2} + s_i \cdot (C1 + C2) & -s_i \cdot C1 & -s_i \cdot C2 \\[2mm]
-s_i \cdot C1 & \dfrac{1}{R3} + s_i \cdot C1 & \dfrac{-1}{R3} \\[2mm]
0 & Ao & 1
\end{bmatrix} \\[6mm]
B \leftarrow \begin{bmatrix} \dfrac{1}{R1} & 0 & 0 \end{bmatrix}^T \\[4mm]
C_i \leftarrow \text{lsolve}(A_i, B) \\[2mm]
Vo_i \leftarrow \left| (C_i)_3 \right| \\[2mm]
Vo_i
\end{vmatrix}
$$

$$Tr := 0.02 \qquad Tc := 0.1 \qquad T := \begin{bmatrix} -Tr & -Tr & -Tr & -Tc & -Tc \\ Tr & Tr & Tr & Tc & Tc \end{bmatrix}$$

$$Nc := \text{cols}(T) \qquad Nc = 5$$

$$p := 1..Nc \qquad dpf := 0.0001 \qquad Q := dpf \cdot \text{identity}(Nc) + 1$$

$$Vo_i := G(R1, R2, R3, C1, C2, s)$$

$$Vr_{i,p} := G(R1 \cdot Q_{p,1}, R2 \cdot Q_{p,2}, R3 \cdot Q_{p,3}, C1 \cdot Q_{p,4}, C2 \cdot Q_{p,5}, s)$$

$$Sen_{i,p} := \left(\frac{Vr_{i,p}}{Vo_i} - 1\right) \cdot \frac{1}{dpf}$$

$$L_{i,p} := if(Sen_{i,p} > 0, 1 + T_{1,p}, 1 + T_{2,p})$$

$$H_{i,p} := if(Sen_{i,p} > 0, 1 + T_{2,p}, 1 + T_{1,p}) \quad k := 1..2$$

$$M_{i,p} := if(k = 1, L_{i,p}, H_{i,p})$$

$$Vrss_{k,i} := Vo_i \cdot \left[1 + (-1)^k \cdot \sqrt{\sum_p \left[Sen_{i,p} \cdot \left(M_{i,p} - 1\right)\right]^2}\right]$$

$$Vev_{1,i} := G(R1 \cdot L_{i,1}, R2 \cdot L_{i,2}, R3 \cdot L_{i,3}, C1 \cdot L_{i,4}, C2 \cdot L_{i,5}, s)$$
EVA low (EVL)

$$Vev_{2,i} := G(R1 \cdot H_{i,1}, R2 \cdot H_{i,2}, R3 \cdot H_{i,3}, C1 \cdot H_{i,4}, C2 \cdot H_{i,5}, s)$$
EVA high (EVH)

Finally, we are able to see the fruits of our efforts. The sensitivities for all five components across the frequency band of interest (400 Hz to 600 Hz) are plotted in Figure 2:

The markers at 488 Hz and 512 Hz, show the approximate peak high and low sensitivities. The sensitivity for R1 is about −1 across the band. We next zoom in on the sensitivities at the crossover frequency fo in Figure 3 to identify the individual sensitivities. The center frequency fo in Hz is given by:

$$fo := \frac{1}{P} \cdot \sqrt{\frac{1}{R3 \cdot C1 \cdot C2} \cdot \left(\frac{1}{R1} + \frac{1}{R2}\right)} \quad P = 6.283 \quad fo = 500.604$$

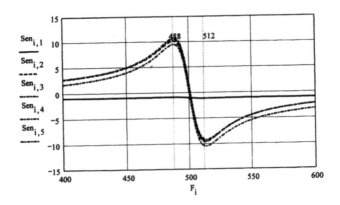

FIGURE 5. Sensitivities.

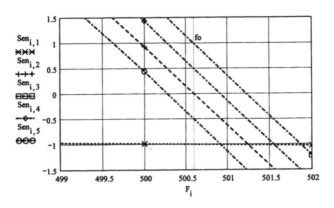

FIGURE 6. Sensitivities near fo.

The sensitivities of the five components at fo are exactly, in R1, R2, R3, C1, C3 order, −1, 0, +1, +0.5, −0.5. These are the active filter design values given in textbooks. Nothing is said about what happens to the sensitivities above and below fo. From Figure 2 we can see that the sensitivities increase dramatically. The consequences of this for the RSS calculations are disastrous as seen in Figure 7.

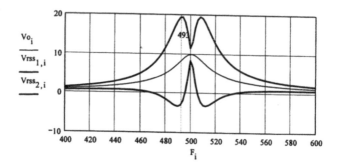

FIGURE 7. RSS.

These results are mathematically correct, but grossly misleading for tolerance analysis purposes. The + 3σ RSS amplitude at 492 Hz is: (counter i = 47 at 493 Hz)

$$F_{47} = 492 \qquad Vo_{47} = 8.23 \qquad Vrss_{2,47} = 19.351$$

To make matters worse, the amplitude is negative at 492 Hz and the units are Volts, not dBV. Hence we have ample reason to suspect that something is amiss. The errors are due to a bad application of a good idea. As shown in the Appendix, the RSS method was not intended to be used for large sensitivity-tolerance (ST) products. The sum of the ST's for C1 and C2 is $(0.1) 10 + (0.1) 10 = 2$, which means we add $\sqrt{2}$ times the nominal Vo. For example:

$$Vo_{47} \cdot \left(1 + \sqrt{2}\right) = 19.868$$

which is about $Vrss_{2,47}$ shown on the previous page.

Again, the mathematics is correct, but the answers are "wrong" in the practical sense. The EVA results, shown in Figure 5 below, are almost as bad. Is it true that at 460 Hz the output will *never* be greater than about 6.6V? As will be shown in Part 2, the answer is **No**. What the plot does show correctly is the min and max excursions of the center frequency of 447 and 568 Hz. No more, no less. Hence it is not a total loss.

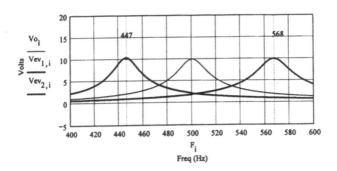

FIGURE 8.   EVA.

SPICE ANALYSIS

While the results of the EVA/RSS analysis for this particular circuit are disappointing, the Spice analysis is more so. Both center frequency excursions are incorrectly shown higher than fo. Shown below is a shortened copy of the *.out file for the same circuit. Note that the error is due to taking the sensitivities at only one frequency (501 Hz below) instead of across the entire frequency band.

```
*************** 01/23/99 08:43:09 ****************
********** Evaluation Spice (July 1997) **********

Bandpass Filter WCA
V1 1 0 1 AC 1
R1 1 2 RA 6.34K
R2 2 0 RA 80.6
R3 4 3 RA 127K
C1 2 3 CA 0.1uF
C2 4 2 CA 0.1uF
E1 4 0 0 3 1E6
* As in MathCad, an ideal VCVS is used.
.MODEL RA RES (R=1 DEV/GAUSS=0.667%)
*3 sigma = 2%
.MODEL CA CAP(C=1 DEV/GAUSS=3.333%)
*3 sigma = 10%
*.WCASE AC V(4) MIN VARY DEV
.WCASE AC V(4) MAX VARY DEV
*.WCASE AC V(4) YMAX VARY DEV
.AC LIN 100 400 600
.PROBE V(4)
.OPTIONS NOECHO NOPAGE
**** SORTED DEVIATIONS OF V(4)
                      TEMPERATURE = 27.000 DEG C

RUN              MAXIMUM VALUE

R3 RA R          10.004 at F = 501.01
                 (    .1536% change per 1% change in
                   Model Parameter)
NOMINAL          10.003 at F = 501.01
C1 CA C          9.9991 at F = 501.01
                 (   -.3444% change per 1% change in
                   Model Parameter)
R2 RA R          9.9942 at F = 501.01
                 (   -.8282% change per 1% change in
                   Model Parameter)
```

```
R1 RA R        9.9925 at F = 501.01
               ( -1.0072% change per 1% change in
                 Model Parameter)
C2 CA C        9.9891 at F = 501.01
               ( -1.3424% change per 1% change in
                 Model Parameter)
```

WORST CASE ALL DEVICES

DEVICE	MODEL	PARAMETER	NEW VALUE
C1	CA	C	.9
C2	CA	C	.9
R1	RA	R	.98
R2	RA	R	.98
R3	RA	R	1.02

The incorrect results of the Spice analysis are shown in Figure 9. The reader is encouraged run the preceding Spice file to verify this. (The Mathcad setup for the following plot is given in the Appendix.)

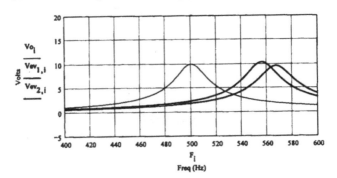

FIGURE 9. Spice EVA.

Comparing Spice and Mathcad sensitivities:

$$fo := 500.6036 \qquad P := 2 \cdot \pi \qquad wo := P \cdot fo \qquad Q := 20$$

The exact C1 sensitivity (derived from the symbolic sensitivity) is:

$$SCl(m) := \frac{m^2 \cdot \left[2 \cdot Q^2 \cdot \left(wo^2 - m^2\right) + wo^2\right]}{2 \cdot \left[Q^2 \cdot \left(wo^2 - m^2\right)^2 + m^2 \cdot wo^2\right]}$$

$m := P \cdot 400, P \cdot 204 .. P \cdot 600$ (m = radian frequency)

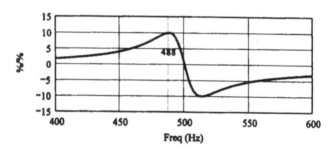

FIGURE 10.   Exact C1 sensitivity.

SC1 (wo) = 0.5; SC1 (P · 488) = 9.991.
Define percent error function:

$$pc\,(a,b) := \frac{a}{b} - 1$$

Percent of sensitivity from Mathcad approximations:

Freq (Hz)	Approximation	Exact	% Error Calculation	% Error
500	$sc1 := 1.4405$	$SC1(P \cdot 500) = 1.461$	$pc1 :=$ $pc(sc1, SC1$ $(P \cdot F_{101}))$	$pc1 = -1.38 \cdot \%$
501	$sc2 := -0.1509$	$SC1(P \cdot 501) = -0.134$	$pc2 :=$ $pc(sc2, SC1$ $(P \cdot F_{102}))$	$pc2 = 12.96 \cdot \%$
502	$sc3 := -1.7144$	$SC1(P \cdot 502) = -1.713$	$pc3 :=$ $pc(sc3, SC1$ $(P \cdot F_{103}))$	$pc3 = 0.06 \cdot \%$

PSpice percent errors from *Spice file:*

Exact	From Spice file	% Error Calculation	% Error
$SC1(P \cdot 501.01) = -0.15$	-0.3444	$pc4 :=$ $pc(-0.3444,$ $SC1(P \cdot 501.01))$	$pc4 = 130.27 \cdot \%$
$SC1(P \cdot 400) = 1.751$	2.7433	$pc5 := pc(2.7433,$ $SC1(P \cdot 400))$	$pc5 = 56.7 \cdot \%$

Hence the Spice user must be aware that the Spice sensitivities may not be as accurate as the Mathcad approximations.

BANDPASS FILTER WITH ASYMMETRIC TOLERANCES

$$T = \begin{bmatrix} -0.02 & -0.02 & -0.02 & -0.15 & -0.15 \\ 0.01 & 0.01 & 0.01 & 0.1 & 0.1 \end{bmatrix}$$

The preceding bandpass filter (BPF) analysis is repeated using the asymmetric tolerances shown in the T array above. For the sake of brevity, only the RSS and EVA plots are shown.

The RSS plot in Figure 11 is asymmetric because the tolerances are. EVL (VL in the plot) in Figure 12 above is increased from 447 Hz in Figure 5 to 451 Hz. EVH (VH) has increased from 568 Hz to about 601 Hz.

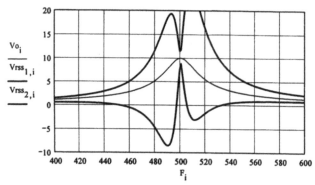

FIGURE 11.   RSS.

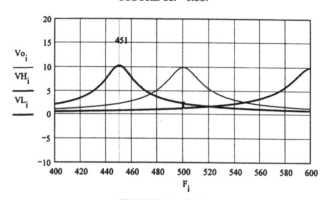

FIGURE 12.   EVA.

LOW-PASS FILTER

So that the reader will not get the impression that ac EVA/RSS analysis always gives wrong or misleading answers, a Butterworth low-pass filter in analyzed using the same methods.

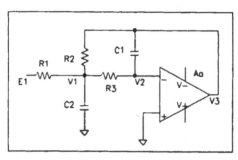

FIGURE 13. LPF sensitivities.

$$K := 10^3 \quad uF := 10^{-6} \quad nF := 10^{-9}$$

$$db(x) := 20 \cdot \log(|x|)$$

$$R1 := 1.43 \cdot K \quad R2 := 14.3 \cdot K$$

$$R3 := 9.09 \cdot K \quad C1 := 2 \cdot nF \quad C2 := 0.1 \cdot uF$$

$$Ao := 10^6$$

$$BF := 2 \quad ND := 2 \quad PD := 50 \quad i := 1 .. ND \cdot PD + 1$$

$$Lg_i := BF + \frac{(i-1)}{PD}$$

$$F_i := 10^{Lg_i} \quad s_i := 2 \cdot \pi \cdot F_i \cdot \sqrt{-1}$$

$$G(R1, R2, R3, C1, C2, s) :=$$

$$
\begin{aligned}
&A_1 \leftarrow
\begin{bmatrix}
\dfrac{1}{R1} + \dfrac{1}{R2} + \dfrac{1}{R3} + s_1 \cdot C2 & \dfrac{-1}{R3} & \dfrac{-1}{R2} \\[2mm]
\dfrac{-1}{R3} & \dfrac{1}{R3} + s_1 \cdot C1 & -s_1 \cdot C1 \\[2mm]
0 & Ao & 1
\end{bmatrix} \\[3mm]
&B \leftarrow \left[\dfrac{1}{R1} \;\; 0 \;\; 0 \right]^T \\[2mm]
&C_1 \leftarrow \text{lsolve}(A_1, B) \\[2mm]
&Vo_1 \leftarrow \left| (C_1)_3 \right| \\[2mm]
&Vo_1
\end{aligned}
$$

$Nc := 5 \qquad p := 1 .. Nc \qquad dpf := 0.0001 \qquad Q := dpf \cdot \text{identity}(Nc) + 1$

$$Vo_i := G(R1, R2, R3, C1, C2, s)$$
$$Vr_{i,p} := G(R1 \cdot Q_{p,1}, R2 \cdot Q_{p,2}, R3 \cdot Q_{p,3}, C1 \cdot Q_{p,4}, C2 \cdot Q_{p,5}, s)$$

$$Sen_{i,p} := \left(\frac{Vr_{i,p}}{Vo_i} - 1 \right) \cdot \frac{1}{dpf} \qquad Tr := 0.02 \quad Tc := 0.1$$

$$T := \begin{bmatrix} -Tr & -Tr & -Tr & -Tc & -Tc \\ Tr & Tr & Tr & Tc & Tc \end{bmatrix}$$

$$L_{i,p} := \text{if}(Sen_{i,p} > 0, \, 1 + T_{1,p}, \, 1 + T_{2,p})$$

$$H_{i,p} := \text{if}(Sen_{i,p} > 0, \, 1 + T_{2,p}, \, 1 + T_{1,p})$$

$$Vrss_{k,i} := Vo_i \cdot \left[\text{if} \left[\begin{array}{l} k = 1, 1 - \sqrt{\sum_{p} \left[Sen_{i,p} \cdot \left(L_{i,p} - 1 \right) \right]^2}, \\ 1 + \sqrt{\sum_{p} \left[Sen_{i,p} \cdot \left(H_{i,p} - 1 \right) \right]^2} \end{array} \right] \right]$$

$$Vev_{1,i} := G\left(R1 \cdot L_{i,1},\ R2 \cdot L_{i,2},\ R3 \cdot L_{i,3},\ C1 \cdot L_{i,4},\ C2 \cdot L_{i,5},\ s\right)$$

$$Vev_{2,i} := G\left(R1 \cdot H_{i,1},\ R2 \cdot H_{i,2},\ R3 \cdot H_{i,3},\ C1 \cdot H_{i,4},\ C2 \cdot H_{i,5},\ s\right)$$

Note log frequency scale below:

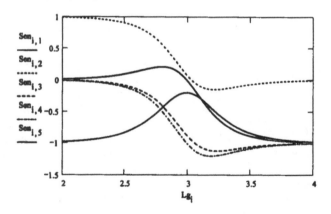

FIGURE 14. LPF sensitivities.

In Figure 14, top to bottom order at 1kHz (Lg = 3) is
R2, C2, R1, R3, C1.
It is important to note that the peak sensitivities for this circuit are
much less than that of the BPF.

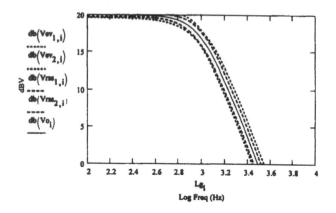

FIGURE 15.   LPF RSS (dashes) and EVA (dots).

This is the output one normally expects; i.e., RSS less than the EVA and symmetric about the nominal (with symmetric tolerances).

Figure 16 expands the plot. The EVA and RSS deviations can now clearly be seen. Note that as some sensitivities approach zero at 1 kHz ($Lg = 3$), the spread gets smaller.

The key to the EVA/RSS behavior of any ac circuit lies in the sensitivities. If the sensitivities are all about one or less, then the ST products will be small and the resulting RSS output will be a better estimate.

The sensitivities should always be computed for possible economic benefit and to better understand the circuit characteristics and behavior. Note that in dc circuits, resistor tolerances seldom are greater than 5% and most ST products will be considerably less.

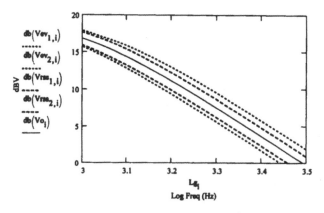

FIGURE 16.   Figure 15 expanded.

2 Monte Carlo Analysis

The general Monte Carlo Analysis (MCA) procedure is to create Nc random numbers, then convert these random numbers to Nc tolerances so that a random tolerances between −Tr and +Tr are generated. These are applied to the Nc components in the circuit and a random output is computed in the same manner as the nominal output was computed.

This is repeated a sufficient number of times (N samples) so that the spread of outputs represents a typical manufacturing run of the circuit. The number of samples N cannot be too large. N = 1000 should be considered a minimum. As N increases, the average (mean) output and standard deviation (σ) statistics tend to stabilize in value.

For ac circuits, N samples of Nc components are taken at say, 200 frequency points. If Nc = 6 and N = 1000, this creates a random number array with 6 columns and 1000 rows at each frequency point. Though in-place storage at each frequency is generally used, ac MCA can tax computer memory.

It should be stated that there are no ideal random number generators. Some come close to being ideal, however virtually all random number algorithms depend in some manner upon the previous random number. A truly random number does not remember the previous one. For example, the next Keno ball in Las Vegas has no knowledge of what the previous ball was, and the two little red cubes at the crap table are similarly heartless and uncooperative.

The following will show how Mathcad creates and uses what should more accurately be called psuedo-random numbers, but will be referred to as random numbers.

Two types of random numbers will be used, Gaussian (normal bell curve) and uniform (flat).

RANDOM NUMBER GENERATOR*

Function to obtain the fractional part of a number:

$$\text{frac}(x) := x - \text{floor}(x)$$

$$N := 20000 \text{ samples.} \qquad \text{seed} := 11$$

A Mathcad program is used to create the random number generator:

$$Rn := \begin{vmatrix} B \leftarrow 0.7654321 \\ \text{for } k \; \varepsilon \; 1..N \\ \quad \begin{vmatrix} C_k \leftarrow \text{frac}(\text{seed} \cdot B) \\ B \leftarrow C_k \end{vmatrix} \\ C \end{vmatrix}$$

$$\text{Mean}(Rn) = 0.499.$$

Histogram Setup:

$$nb := 20 \text{ (Number of bins)} \qquad q := 1..nb + 1 \qquad nh := 1..nb$$

$$H := \max(Rn) \qquad\qquad L := \min(Rn) \qquad intv := \frac{H - L}{nb}$$

$$bin_q := L + intv \cdot (q - 1) \qquad pu := \text{hist}(bin, Rn)$$

---

* Algorithms for RPN Calculators, Ball, Wiley, 1978, p. 206.

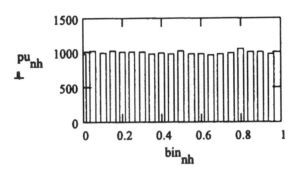

FIGURE 17. Bar histogram. Note the even (uniform) distribution of random numbers from 0 to 1.0.

MATHCAD'S RANDOM NUMBER GENERATOR

$k := 1 .. N$ $\qquad$ $Rm_k := rnd(1)$ $\qquad$ $mean(Rm) = 0.502$

$H := max(Rm)$ $\qquad$ $L := min(Rm)$ $\qquad$ $intv := \dfrac{H - L}{nb}$

$bin_q := L + intv \cdot (q - 1)$ $\qquad$ $pm := hist(bin, Rm)$

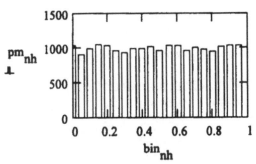

FIGURE 18. Bar histogram.

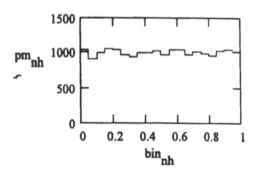

FIGURE 19. Step histogram.

GAUSSIAN RANDOM NUMBER GENERATOR (RNG)

The Mathcad morm function is used to create 6 random numbers for 4 components for a total of 24:

$N := 6$ (No. of samples) $Nc := 4$ (No. of components)

$k := 1 .. N$ $w := 1 .. Nc$ $z_w := morm(N, 0, 1)$

         (NxNc Gaussian random numbers with 0 mean and a standard deviation (σ) of 1.)

$za := augment(z_1, augment(z_2, augment(z_3, z_4)))$

$$za = \begin{bmatrix} -0.439 & -0.121 & 0.916 & -0.182 \\ -0.679 & 0.556 & 0.673 & -0.644 \\ -0.473 & 2.192 & -1.044 & -0.723 \\ -0.951 & 0.809 & 0.069 & -0.517 \\ -1.686 & 0.985 & -0.756 & 0.558 \\ 0.044 & 0.862 & 0.697 & -0.245 \end{bmatrix}$$

99.9% of the za numbers will be within -3 & $+3$.

CONVERTING TO RANDOM TOLERANCES

$x := -3, -2.5 .. 3$ (Limit the input to +/−3σ.)

$T1(x) := \dfrac{0.02 + 0.02}{6} \cdot (x + 3) - 0.02$ (Symmetric tolerances of +/−2%)

$T2(x) := \dfrac{0.03 + 0.01}{6} \cdot (x + 3) - 0.01$ (Asymmetric tolerances of +3% and −1%)

$T3(x) := \dfrac{0.01 + 0.03}{6} \cdot (x + 3) - 0.03$ (Asymmetric tolerances of +1% and −3%)

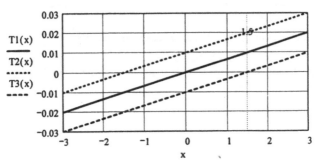

FIGURE 20. Plot of T1, T2, and T3.

In Figure 20, at $x = +1.5$, the tolerances are:
$T1(1.5) = 1 \cdot \%$, $T2(1.5) = 2 \cdot \%$, and $T3(1.5) = 0 \cdot \%$.

Next, get $N = 6$ random tolerances of 5% for $Nc = 4$ components based on z above $Tr := 0.05$

Tolerance array T:

$$T := \begin{bmatrix} -Tr & -Tr & -Tr & -Tr \\ Tr & Tr & Tr & Tr \end{bmatrix}$$

$$Tn_{k,w} := \frac{T_{2,w} - T_{1,w}}{6} \cdot \left[(z_w)_k + 3 \right] + T_{1,w} + 1$$

$$Tn = \begin{bmatrix} 0.993 & 0.998 & 1.015 & 0.997 \\ 0.989 & 1.009 & 1.011 & 0.989 \\ 0.992 & 1.037 & 0.983 & 0.988 \\ 0.984 & 1.013 & 1.001 & 0.991 \\ 0.972 & 1.016 & 0.987 & 1.009 \\ 1.001 & 1.014 & 1.012 & 0.996 \end{bmatrix}$$

Each component (column) has a unique random tolerance *multiplier* between 0.95 and 1.05.

Each number in $Tn > 1$ corresponds to za numbers > 0.

Pick out one random tolerance: $Tn_{5,3} = 0.987$. Hence if the nominal component value were 1000Ω, it would be changed to 987Ω when multiplied by $Tn_{5,3}$.

Histogram of Gaussian RNG:

$$N := 10000 \qquad\qquad z := \text{rnorm}(N, 0, 1) \qquad nb := 30$$

$$q := 1 .. nb + 1 \qquad\qquad nh := 1 .. nb \qquad\qquad VLg := \min(z)$$

$$VHg := \max(z) \qquad\qquad intv := \frac{VHg - VLg}{nb}$$

$$bin_q := VLg + intv \cdot (q - 1) \qquad\qquad\qquad pv := \text{hist}(bin, z)$$

Plot ideal Gaussian curve to compare with histogram:

$$E(z) := N \cdot intv \cdot \text{dnorm}(z, 0, 1)$$

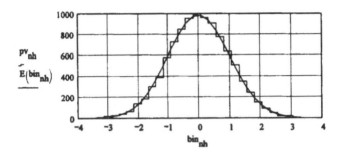

FIGURE 21. Gaussian curve and histogram. Max(pv) = 993.

Histogram of *uniform* RNG:

$$v := \text{runif}(N, -3, 3) \qquad \text{(\textit{runif} is a uniform RNG;}$$
$$\textit{rnorm is a Gaussian RNG)}$$

$$VLu := \min(v) \qquad VHu := \max(v) \qquad \text{intv} := \frac{VHu - VLu}{nb}$$

$$\text{bin}_q := VLu + \text{intv} \cdot (q - 1) \qquad\qquad pu := \text{hist}(\text{bin}, v)$$

$$x := -4, -3.92 .. 4 \qquad U(x) := \frac{N}{nb} \cdot (\phi(x+3) - \phi(x-3))$$

(U is an ideal continuous uniform distribution)

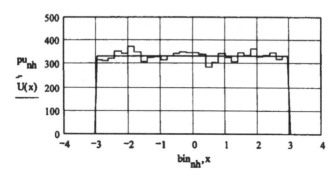

FIGURE 22. Uniform distribution. Max (pu) = 374.

Compare the extremes (min and max) of the two distributions:

Gaussian: VLg = –3.885   VHg = 3.546
Uniform: VLu = –3   VHu = 3

Hence components multiplied by VLg or VHg would "out of spec" and the simulation would be somewhat unrealistic. (These extremes are called "outliers" in statistics.) But in the uniform distribution, there would be no "out of spec" components used in the simulation. Thus using Gaussian distribution may be undesireable in some cases, such as in estimating manufacturing yield.

MONTE CARLO ANALYSIS — DC DIFFERENTIAL AMPLIFIER

We next perform a MCA on the differential amplifier and compare the results to the previous RSS/EVA analysis of that same circuit.

E1 := 1   E2 := –1   R1 := 10   R2 := 100   R3 := 10

$$R4 := 100 \qquad G(R1, R2, R3, R4, E1, E2) := \frac{E1 \cdot \left(1 + \dfrac{R2}{R1}\right)}{1 + \dfrac{R3}{R4}} - \frac{E2 \cdot R2}{R1}$$

$$Vo := G(R1, R2, R3, R4, E1, E2)$$

Begin MCA:

$$N := 20000 \qquad Nc := 6 \qquad k := 1..N \qquad w := 1..Nc$$
$$z_w := rnorm(N, 0, 1)$$

Tolerance array T:

$$Tr := 0.01 \qquad Te := 0.05 \qquad T := \begin{bmatrix} -Tr & -Tr & -Tr & -Tr & -Te & -Te \\ Tr & Tr & Tr & Tr & Te & Te \end{bmatrix}$$

Convert to tolerance multipliers:

$$Tn_{k,w} := \frac{\left(T_{2,w} - T_{1,w}\right)}{6} \cdot \left[\left(z_w\right)_k + 3\right] + T_{1,w} + 1$$

Insert the random tolerance multipliers into the G-function to get N random outputs Vm:

$$Vm_k := G\begin{bmatrix} (R1) \cdot Tn_{k,1}, R2 \cdot Tn_{k,2}, R3 \cdot Tn_{k,3}, R4 \cdot Tn_{k,4}, \\ E1 \cdot Tn_{k,5}, E2 \cdot Tn_{k,6} \end{bmatrix}$$

Get statistics:

$$Vavg := mean(Vm) \qquad Vs := stdev(Vm) \qquad VL := min(Vm)$$
$$VH := max(Vm)$$

Statistical confidence intervals: (see "Advanced Topics")

Histogram:

$$nb := 40 \qquad q := 1..nb + 1 \qquad nh := 1..nb \qquad intv := \frac{VH - VL}{nb}$$

$$bin_q := VL + intv \cdot (q - 1) \qquad\qquad pv := hist\,(bin, Vm)$$

$$E\,(Vm := N \cdot intv \cdot dnorm\,(Vm, Vavg, Vs) \qquad n := 1..2$$

$$Vmca_n := Vavg + (-1)^n \cdot 3 \cdot Vs \qquad\qquad \Delta Vmca := Vavg - Vmca$$

$$Vmca = \begin{bmatrix} 19.241 \\ 20.76 \end{bmatrix} \qquad \Delta Vmca = \begin{bmatrix} 0.759 \\ -0.759 \end{bmatrix} \qquad Vo = 20$$

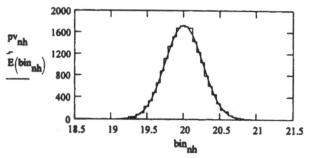

FIGURE 23. Symmetric tolerances.

Now use asymmetric tolerances as in Part 1 example (see page 22).

$$T := \begin{bmatrix} -Tr & -Tr & -Tr & -Tr & -Te & -Te \\ 0.03 & Tr & 0.03 & Tr & Te & Te \end{bmatrix}$$

$$Tn_{k,w} := \frac{(T_{2,w} - T_{1,w})}{6} \cdot \left[(z_w)_k + 3\right] + T_{1,w} + 1$$

$$Vn_k := G \begin{bmatrix} (R1) \cdot Tn_{k,1}, R2 \cdot Tn_{k,2}, R3 \cdot Tn_{k,3}, R4 \cdot Tn_{k,4}, \\ E1 \cdot Tn_{k,5}, E2 \cdot Tn_{k,6} \end{bmatrix}$$

$$Vavg := mean(Vn) \qquad VL := min(Vn) \qquad VH := max(Vn)$$

$$Vs := stdev(Vn)$$

$$intv := \frac{VH - VL}{nb} \qquad bin_q := VL + intv \cdot (q - 1)$$

$$pa := hist(bin, Vn) \qquad E(Vn) := N \cdot intv \cdot dnorm(Vn, Vavg, Vs)$$

$$Vmca_n := Vavg + (-1)^n \cdot 3 \cdot Vs \qquad \Delta Vmca := Vavg - Vmca$$

$$Vmca = \begin{bmatrix} 18.983 \\ 20.623 \end{bmatrix} \qquad \Delta Vmca = \begin{bmatrix} 0.82 \\ -0.82 \end{bmatrix} \qquad Vavg = 19.803$$

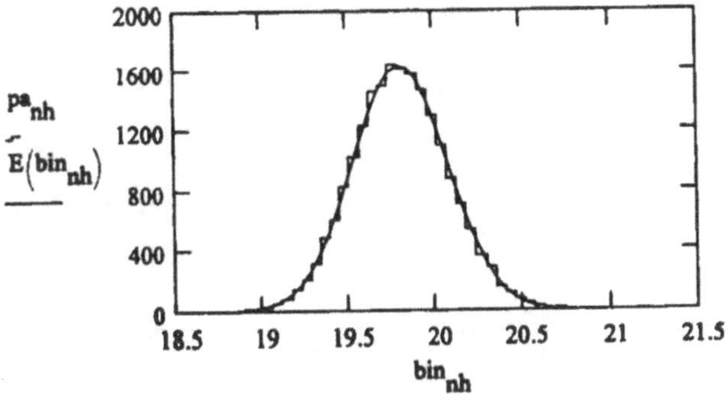

FIGURE 24. Asymmetric tolerances.

Results from page 24:

$$\text{Vrss} := \begin{bmatrix} 19.07 \\ 20.757 \end{bmatrix} \qquad \overrightarrow{\frac{\text{Vrss}}{\text{Vmca}}} - 1 = \begin{bmatrix} 0.446 \\ 0.692 \end{bmatrix} \cdot \%$$

% error of RSS vs. MCA

Less than 1%. Hence this gives credence to the RSS asymmetric tolerance method introduced earlier.

UNIFORM DISTRIBUTION INPUT

There are two ways to do this:

(1) $\quad \text{Tn}_{k,w} := (T_{2,w} - T_{1,w}) \cdot \text{rnd}(1) + T_{1,w} + 1$

or

(2) $\quad v_w := \text{runif}(N, -3, 3)$

$$\text{Tn}_{k,w} := \frac{(T_{2,w} - T_{1,w})}{6} \cdot \left[(v_w)_k + 3 \right] + T_{1,w} + 1$$

Method (1) is faster and therefore preferred.

$$\text{Vm}_k := G \begin{bmatrix} (R1) \cdot \text{Tn}_{k,1}, R2 \cdot \text{Tn}_{k,2}, R3 \cdot \text{Tn}_{k,3}, R4 \cdot \text{Tn}_{k,4}, \\ E1 \cdot \text{Tn}_{k,5}, E2 \cdot \text{Tn}_{k,6} \end{bmatrix}$$

$\text{Vavg} := \text{mean}(\text{Vm}) \qquad \text{VL} := \min(\text{Vm}) \qquad \text{VH} := \max(\text{Vm})$

$\text{Va} := \text{var}(\text{Vm})$

$\text{intv} := \dfrac{\text{VH} - \text{VL}}{\text{nb}} \quad \text{bin}_q := \text{VL} + \text{intv} \cdot (q-1) \quad \text{pa} := \text{hist}(\text{bin}, \text{Vm})$

$E(\text{Vm}) := N \cdot \text{intv} \cdot \text{dnorm}\left(\text{Vm}, \text{Vavg}, \sqrt{\text{Va}}\right)$

$\text{Vmcu}_a := \text{Vavg} + (-1)^a \cdot 3 \cdot \sqrt{\text{Va}} \qquad \Delta\text{Vmcu} := \text{Vavg} - \text{Vmcu}$

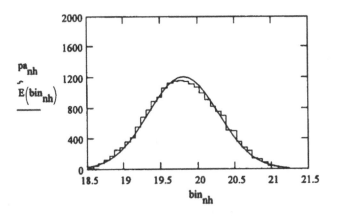

FIGURE 25. Uniform input. Same scale as Figure 24.

$$\text{Vmcu} = \begin{bmatrix} 18.39 \\ 21.22 \end{bmatrix} \qquad \Delta\text{Vmcu} = \begin{bmatrix} 1.415 \\ -1.415 \end{bmatrix}$$

For a uniform input: *if the output is Gaussian* then the 3σ points are $\sqrt{3}$ times the 3s points for a corresponding Gaussian input/output.

Vmca used a Gaussian input: then the Δ's for the uniform distribution are:

$$\sqrt{3} \cdot \Delta\text{Vmca} = \begin{bmatrix} 1.411 \\ -1.411 \end{bmatrix}$$

which is approximately ΔVmcu above. See Appendix for the derivation.

MCA OF RTD CIRCUIT

(For a schematic, component values, tolerances, and matrix development see pages 36 to 38.)

From previous work: $Vo = 4.326$

$$T := \begin{bmatrix} Tlo & Tlo & Tlo & Tlo & Tlo & Tlo & Tlo & Tlo & Tlo & Trlo & Treflo \\ Thi & Thi & Thi & Thi & Thi & Thi & Thi & Thi & Thi & Trhi & Trefhi \end{bmatrix}$$

$N := 10000$ $Nc := 11$ $k := 1..N$ $w := 1..Nc$

$$z_w := \text{rnorm}(N, 0, 1) \quad Tn_{k,w} := \frac{(T_{2,w} - T_{1,w})}{6} \cdot \left[(z_w)_k + 3\right] + T_{1,w} + 1$$

$$Vm_k := G\begin{pmatrix} R1 \cdot Tn_{k,1}, R2 \cdot Tn_{k,2}, R3 \cdot Tn_{k,3}, R4 \cdot Tn_{k,4}, R5 \cdot Tn_{k,5}, \\ R6 \cdot Tn_{k,6}, R7 \cdot Tn_{k,7}, R8 \cdot Tn_{k,8}, R9 \cdot Tn_{k,9}, RT \cdot Tn_{k,10}, \\ E1 \cdot Tn_{k,11} \end{pmatrix}$$

$Vavg := \text{mean}(Vm)$ $VL := \text{min}(Vm)$ $VH := \text{max}(Vm)$

$nb := 30$ $q := 1..nb + 1$ $nh := 1..nb$

$$intv := \frac{VH - VL}{nb} \qquad m := 1..2 \qquad bin_q := VL + intv \cdot (q - 1)$$

$$pv := \text{hist}(bin, Vm) \qquad Vg := \text{stdev}(Vm)$$

$$E(Vm) := N \cdot intv \cdot \text{dnorm}(Vm, Vavg, Vg)$$

$$Vmca_m := Vavg + (-1)^m \cdot 3 \cdot Vg$$

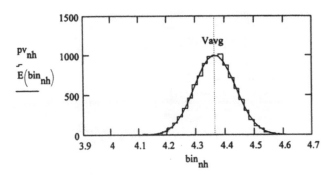

FIGURE 26. RTD circuit.

$$Vavg = 4.365 \qquad Vo = 4.326 \qquad Vmca = \begin{bmatrix} 4.157 \\ 4.574 \end{bmatrix}$$

From previous work: (page 40) $\quad Vrss := \begin{bmatrix} 4.122 \\ 4.55 \end{bmatrix}$

Comparing MCA numbers with RSS numbers:

$$\overrightarrow{\frac{Vrss}{Vmca}} - 1 = \begin{bmatrix} -0.841 \\ -0.519 \end{bmatrix} \cdot \% \qquad \text{(Less than 1\% error)}$$

Uniform input:

$$Tn_{k,w} := (T_{2,w} - T_{1,w}) \cdot rnd(1) + T_{1,w} + 1$$

$$Vu_k := G \begin{pmatrix} R1 \cdot Tn_{k,1}, R2 \cdot Tn_{k,2}, R3 \cdot Tn_{k,3}, R4 \cdot Tn_{k,4}, R5 \cdot Tn_{k,5}, \\ R6 \cdot Tn_{k,6}, R7 \cdot Tn_{k,7}, R8 \cdot Tn_{k,8}, R9 \cdot Tn_{k,9}, RT \cdot Tn_{k,10}, \\ E1 \cdot Tn_{k,11} \end{pmatrix}$$

$$\text{Vavg} := \text{mean}(\text{Vu}) \qquad \text{VL} := \min(\text{Vu}) \qquad \text{VH} := \max(\text{Vu})$$

$$\text{intv} := \frac{\text{VH} - \text{VL}}{\text{nb}} \qquad \text{bin}_q := \text{VL} + \text{intv} \cdot (q - 1)$$

$$\text{pv} := \text{hist}(\text{bin}, \text{Vu})$$

$$\text{Vs} := \text{stdev}(\text{Vu}) \qquad E(\text{Vu}) := N \cdot \text{intv} \cdot \text{dnorm}(\text{Vu}, \text{Vavg}, \text{Vs})$$

$$\text{Vmca}_m := \text{Vavg} + (-1)^m \cdot 3 \cdot \text{Vs} \qquad \text{Vmca} = \begin{bmatrix} 4.006 \\ 4.728 \end{bmatrix}$$

From previous work:

$$\text{Vrsu} = \begin{bmatrix} 3.972 \\ 4.714 \end{bmatrix} \qquad \overrightarrow{\frac{\text{Vrsu}}{\text{Vmca}}} - 1 = \begin{bmatrix} -0.859 \\ -0.302 \end{bmatrix} \cdot \% \qquad \begin{matrix} \text{(Less than} \\ \text{1\% error.)} \end{matrix}$$

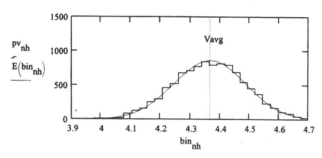

FIGURE 27.   RTD output.

Check $\sqrt{3}$ factor:

$$\text{Vg} = 0.0695 \qquad \text{(Normal dist. } \sigma\text{)}$$

$$\sqrt{3} \cdot \text{Vg} = 0.1203 \qquad \text{(Approx = uniform dist. } \sigma\text{)}$$

$$Vs = 0.1203 \qquad \text{(Uniform dist. } \sigma\text{)}$$

$$\frac{\sqrt{3} \cdot Vg}{Vs} - 1 = 0.007\% \qquad \text{(Percent error.)}$$

The percent error here indicates that the output is very close to Gaussian. The more non-Gaussian the output, the higher this error will be.

MCA BANDPASS FILTER (BPF)

We next perform a Monte Carlo analysis on the BPF of Part 1 using the transfer function instead of matrix analysis. If a simple transfer function is available, it should be used as it can lead to more insight into circuit behavior. The 2nd order BPF transfer function is:

$$F(s) = \frac{N1 \cdot s}{s^2 + D1 \cdot s + D0}$$

The N1, D1, and D0 coefficient are given in the Mathcad program below:

$$K := 10^3 \qquad uF := 10^{-6} \qquad j := \sqrt{-1} \qquad R1 := 6.34 \cdot K$$

$$R2 := 80.6 \qquad R3 := 127 \cdot K \qquad C1 := 0.1 \cdot uF \qquad C2 := C1$$

$$Bf := 400 \qquad LF := 600 \qquad DF := 1 \qquad lit := \frac{LF - BF}{DF} + 1$$

$$i := 1 .. lit \qquad F_i := BF + DF \cdot (i - 1) \qquad s_i := 2 \cdot \pi \cdot F_i \cdot j$$

$$N := 5 \qquad Nc := 5 \qquad Tr := 0.02 \qquad Tc := 0.1$$

$$T := \begin{bmatrix} -Tr & -Tr & -Tr & -Tc & -Tc \\ Tr & Tr & Tr & Tc & Tc \end{bmatrix}$$

$$G(R1, R2, R3, C1, C2, s) := \begin{vmatrix} N1 \leftarrow \dfrac{1}{R1 \cdot C1} \\ D1 \leftarrow \dfrac{1}{R3} \cdot \left(\dfrac{1}{C1} + \dfrac{1}{C2} \right) \\ D0 \leftarrow \dfrac{1}{R3 \cdot C1 \cdot C2} \left(\dfrac{1}{R1} + \dfrac{1}{R2} \right) \\ Vo \leftarrow \left| \dfrac{N1 \cdot s_i}{(s_i)^2 + D1 \cdot s_i + D0} \right| \\ Vo \end{vmatrix}$$

$$Vo_i := G(R1, R2, R3, C1, C3, s) \qquad \text{(Nominal output)}$$

$$fo := 500.6 \qquad w := 1 .. Nc \qquad k := 1 .. N \qquad z_w := rnorm(N, 0, 1)$$

$$Tn_{k,w} := \frac{T_{2w} - T_{1,w}}{6} \cdot \left[(z_w)_k + 3 \right] + T_{1,w} + 1$$

Multiply components in the G-function by the random tolerance multipliers Tn:

$$Vm_{k,i} := G\left(R1 \cdot Tn_{k,1}, R2 \cdot Tn_{k,2}, R3 \cdot Tn_{k,3}, C1 \cdot Tn_{k,4}, C2 \cdot Tn_{k,5}, s\right)$$

$$Vmin_i := min(Vm^{<i>}) \qquad Vmax_i := max(Vm^{<i>})$$

(Get the min and max at each frequency)

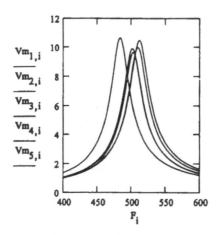

FIGURE 28.   MCA - five samples of the BPF.

The plot shows the five samples of the BPF. Normally a minimum of 1000 samples will be taken to get more accurate statistics. Hence 1000 traces would result in an illegible smear.

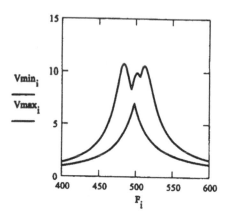

FIGURE 29   MCA - min/max of all five samples of the BPF.

The plot draws only the minimums and maximums of all five samples and shows the output spread in a compact form. The minimums are the right and left-hand skirts of the outermost plots, and hence are of less interest. This is the output format that will be used for ac MCA.

Next we use a sample size of $N := 1000 \qquad k := 1 .. N$

$$z_w := \text{morm}(N, 0, 1)$$

$$Tn_{k,w} := \frac{T_{2w} - T_{1,w}}{6} \cdot \left[(z_w)_k + 3 \right] + T_{1,w} + 1$$

$$Vm_{k,i} := G(R1 \cdot Tn_{k,1}, R2 \cdot Tn_{k,2}, R3 \cdot Tn_{k,3}, C1 \cdot Tn_{k,4}, C2 \cdot Tn_{k,5}, s)$$

$$Vmin_i := \min(Vm^{<i>}) \qquad Vmax_i := \max(Vm^{<i>})$$

$$Vpk := \max(Vmax) \qquad Vpk = 10.761$$

Vpk is the "maximum of the maximums." The BPF MCA plot is shown on the following page. Note how the G-function concept produces very compact and efficient Mathcad statements. We can multiply the components R1, ... C2 by virtually any parameter when using this method.

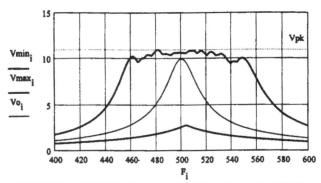

FIGURE 30. BPF MCA (heavy) and Vo (light) plot. Vpk = 10.961

Hence we can see the peak amplitude Vpk. Increasing the number of samples N would cause the min and max center frequencies to shift farther out from Fo.

We next plot the Vavg + 3σ value and compare with the BPF MCA plot.

$$Vs_i := stdev(Vm^{<i>}) \qquad Vavg_i := mean(Vm^{<i>})$$
$$V1_i := Vavg_i + 3 \cdot Vs_i \qquad V1pk := max(V1)$$
$$V1pk = 13,457 \qquad Vpk = 10.961$$

FIGURE 31. Vo and Vavg + 3s.

Note the difference in the maximum 3σ value (V1pk) compared to the maximum of the MCA (Vpk). We have seen that the MCA plot represents the absolute maximums of the BPF output. How can the 3σ value be larger than this? The answer is that at each frequency increment F_i, the outputs are extremely non-Gaussian and hence the 3σ value becomes almost meaningless. We can only visualize where the 3σ points are on the distribution when it is a bell curve or Gaussian shape. For other distributions the 3σ location will be unique to that distribution. The following matrix and 3-D bar histogram will illustrate this.

The following matrix shows the number of samples in each amplitude bin for every 10 Hz of frequency change. Column 1 shows the frequency, while the last row shows the amplitude of each bin in volts. Each row should sum to $N = 2000$ samples. Note that at each frequency the distributions are indeed non-Gaussian.

$$_{rev}^{T} =$$

400	386	1614	0	0	0	0	0	0	0	0	0	0	0	0	0
410	66	1933	1	0	0	0	0	0	0	0	0	0	0	0	0
420	8	1963	29	0	0	0	0	0	0	0	0	0	0	0	0
430	0	1778	221	1	0	0	0	0	0	0	0	0	0	0	0
440	0	1207	750	42	1	0	0	0	0	0	0	0	0	0	0
450	0	524	1209	224	37	5	0	1	0	0	0	0	0	0	0
460	0	127	1043	572	173	49	24	7	4	1	0	0	0	0	0
470	0	21	489	682	385	205	101	55	25	33	4	0	0	0	0
480	0	3	122	415	426	317	222	174	132	153	36	0	0	0	0
490	0	0	21	119	254	251	266	269	284	396	140	0	0	0	0
500	0	0	4	29	117	180	257	305	337	599	172	0	0	0	0
510	0	0	10	95	200	273	262	258	324	456	122	0	0	0	0
520	0	0	77	334	396	340	248	180	168	205	52	0	0	0	0
530	0	4	331	634	443	237	167	66	53	54	11	0	0	0	0
540	0	49	813	669	292	84	49	20	15	7	2	0	0	0	0
550	0	232	1202	428	98	26	8	2	3	0	1	0	0	0	0
560	0	646	1177	148	23	3	2	1	0	0	0	0	0	0	0
570	0	1232	727	36	4	1	0	0	0	0	0	0	0	0	0
580	0	1670	323	6	1	0	0	0	0	0	0	0	0	0	0
590	0	1904	95	1	0	0	0	0	0	0	0	0	0	0	0
600	5	1972	23	0	0	0	0	0	0	0	0	0	0	0	0
0	1	2	3	4	5	6	7	8	9	10	11	12	13	14	15

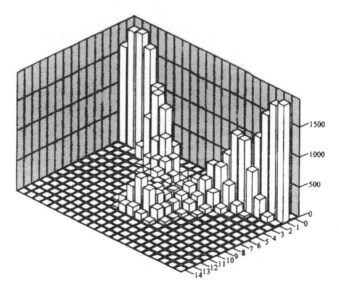

FIGURE 32.   3-D histogram — BPF.

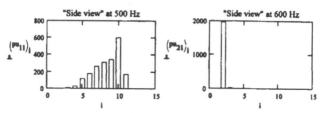

FIGURE 33.   Side view at 500 and 600 Hz.

FAST MONTE CARLO ANALYSIS (FMCA)

To illustrate an inherent weakness in EVA using signed sensitivities, another approach to the EVA problem is shown. In this method, all possible tolerance combinations are used. To see how this is implemented in Mathcad, suppose one were to throw three coins into the air: a quarter, dime, and nickel. What are all the possible combinations of heads and tails for the separate coins? We have TTT, TTH, THT, THH, ... etc. The reader might recognize that a 3-bit binary counter is being formed, with $T = 0$ and $H = 1$. Then to get all possible min and max tolerance combinations for a circuit with three components, there are $2^3 = 8$ possible combinations. So we create a binary counter and use the low tolerance for a zero and the high tolerance for a one.

We start with a simple 3-component voltage divider as given on page 1.2:

$$E1 := 10 \quad K := 10^3 \quad R1 := 10 \cdot K \quad R2 := 30 \cdot K$$

$$G(R1, R2, E1) := \frac{E1 \cdot R2}{R1 + R2}$$

$$Vo := G(R1, R2, E1) \quad Vo = 7.5 \quad Tr := 0.02 \quad Te := 0.05$$

$$T := \begin{bmatrix} -Tr & -Tr & -Te \\ Tr & Tr & Te \end{bmatrix}$$

3-bit binary counter:

$$Nc := 3 \quad N := 2^{Nc} \quad k := 1..N \quad w := 1..Nc \quad Re_{w,k} := k$$

$$Re_{w+1,k} := floor\left(\frac{Re_{w,k}}{2}\right) \quad Dr_{w,k} := Re_{w,k} - 2 \cdot Re_{w+1,k}$$

$$Dr = \begin{bmatrix} 1 & 0 & 1 & 0 & 1 & 0 & 1 & 0 \\ 0 & 1 & 1 & 0 & 0 & 1 & 1 & 0 \\ 0 & 0 & 0 & 1 & 1 & 1 & 1 & 0 \end{bmatrix}$$

Use the low tolerance for binary 0 and the high tolerance for binary 1:

$$Tn_{w,k} := if(Dr_{w,k} = 0, 1 + T_{1,w}, 1 + T_{2,w})$$

$$Tn = \begin{bmatrix} 1.02 & 0.98 & 1.02 & 0.98 & 1.02 & 0.98 & 1.02 & 0.98 \\ 0.98 & 1.02 & 1.02 & 0.98 & 0.98 & 1.02 & 1.02 & 0.98 \\ 0.95 & 0.95 & 0.95 & 1.05 & 1.05 & 1.05 & 1.05 & 0.95 \end{bmatrix}$$

Row order: R1, R2, E1

Then there are 8 different outputs with the eight combinations:

$$Vm_k := G(R1 \cdot Tn_{1,k}, R2 \cdot Tn_{2,k}, E1 \cdot Tn_{3,k})$$

$$VmT = [7.053 \ 7.196 \ 7.125 \ 7.875 \ 7.795 \ 7.953 \ 7.875 \ 7.125]$$

For only 8 outputs, it is easy to pick out the minimum and maximum. But we let Mathcad do it here and for larger circuits.

$$VL := min(Vm) \qquad VH := max(Vm) \qquad VL = 7.053 \quad VH = 7.953$$

These are the same answers given by the first EVA method. The inherent weakness becomes apparent for ac EVA as illustrated by the analysis given on the next page.

The name "Fast Monte Carlo" is coined because the method ignores all intermediate tolerances and immediately chooses the extreme tolerances.

FMCA – BANDPASS FILTER

$$K := 10^3 \qquad uF := 10^{-6} \qquad P := 2 \cdot \pi$$

$$R1 := 6.34K \qquad R2 := 80.6 \qquad R3 := 127 \cdot K$$

$$C1 := 0.1 \cdot uF \qquad C2 := C1$$

$$BF := 400 \qquad LF := 600 \qquad DF := 1$$

$$\text{lit} := \frac{LF - BF}{DF} + 1 \qquad i := 1 .. \text{lit} \qquad F_i := BF + DF \cdot (i - 1)$$

$$s_i := P \cdot F_i \cdot \sqrt{-1} \qquad Nc := 5 \qquad N := 2^{Nc} \qquad k := 1 .. N$$

$$w := 1 .. Nc \qquad Re_{k, w} := k$$

There are 5 components, so we have 32 possible tolerance combinations. $N = 32$

$$Tr := 0.02 \qquad T3 := 0.1 \qquad T := \begin{bmatrix} -Tr & -Tr & -Tr & -Tc & -Tc \\ Tr & Tr & Tr & Tc & Tc \end{bmatrix}$$

$$Re_{k, w+1} := \text{floor}\left(\frac{Re_{k, w}}{2}\right) \qquad Dr_{k, w} := Re_{k, w} - 2 \cdot Re_{k, w+1}$$

$$Tn_{k, w} := \text{if}(Dr_{k, w} = 0, 1 + T_{1, w}, 1 + T_{2, w}) \quad \text{(Tolerance array with min/max tolerances)}$$

BPF transfer function used instead of matrix analysis:

$$G(R1, R2, R3, C1, C2, s) :=$$

$$\frac{\dfrac{s_i}{R1 \cdot C1}}{(s_i)^2 + \dfrac{s_i}{R3} \cdot \left(\dfrac{1}{C1} + \dfrac{1}{C2}\right) + \dfrac{1}{R3 \cdot C1 \cdot C2} \cdot \left(\dfrac{1}{R1} + \dfrac{1}{R2}\right)}$$

$$Vo_i := |G(R1, R2, R3, C1, C2, s)| \qquad \text{(Nominal output)}$$

$$Vm_{k, i} := |G(R1 \cdot Tn_{k, 1}, R2 \cdot Tn_{k, 2}, R3 \cdot Tn_{k, 3}, C1 \cdot Tn_{k, 4}, C2 \cdot Tn_{k, 5}, s)|$$

$$Vmax_i := \max(Vm^{\langle i \rangle}) \qquad Vpk := \max(Vmax) \qquad Vpk = 11.465$$

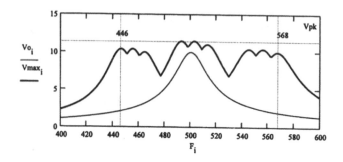

FIGURE 34.   FMCA of BPF.

The FMCA plot shows that 9 of the 32 tolerance combinations result in a peak amplitude at different center frequencies. The EVA plot of Figure 8 did indeed leave out much information and therefore is of little value. The markers show the min and max center frequencies from Figure 8. The FMCA plot shows more worst-case information, but it is *still incomplete* as will be seen.

MCA OF BPF

For the sake of brevity, only the results of the MCA for the BPF are shown. The Mathcad sequence is given in the Appendix.

Two plots are given, one with N = 100 samples (Figure 31), and the second with N = 1000 samples (Figure 36).

$$Vpk = 10.731 \qquad Vfpk = 11.465 \qquad N = 100$$

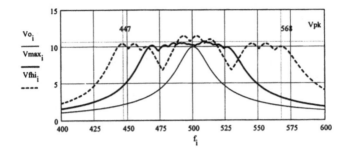

Figure 35. FMCA (dash) and MCA (solid).

Shown below is the same plot but with N = 1000 samples.

$$Vpk = 10.961 \qquad Vfpk = 11.465 \qquad N = 1000$$

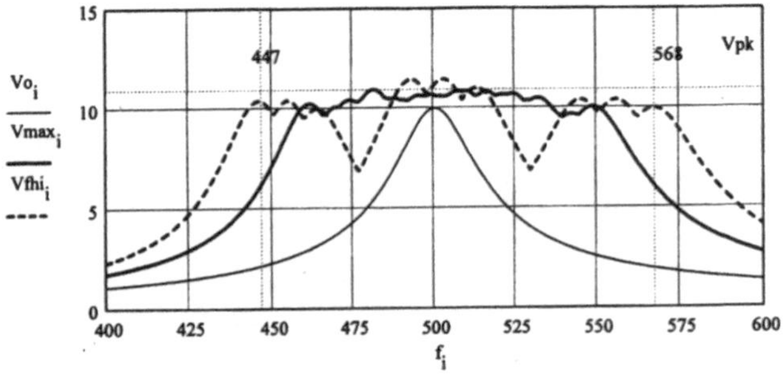

FIGURE 36. FMCA and MCA.

Note the difference between the MCA peak (Vpk) and the absolute max FMCA peak (Vfpk) in both cases. Also note how the MCA band edges approach the extreme center frequencies 447 Hz and 568 Hz with increasing N.

With regard to the interior valleys of the FMCA at about 477 Hz and 530 Hz, these points represent some combination of extreme tolerances. The MCA plot is well above these valleys with some random combination of tolerances *less than* the extreme tolerances. Hence, *output magnitudes can be greater at smaller tolerances*. This is a subtle characteristic of the so-called extreme value analyses and will be explored further in the following pages.

The reader can now understand that a *true* EVA, for this ac circuit and by implication others, can only be obtained with a MCA using an infinite number of samples N. In other words, a true EVA does not exist. As in most statistical analyses, a true EVA value can only be estimated. The accuracy of the estimation improves as N is increased.

COMPONENT SLOPES

A component slope is the slope of the circuit output versus the component value. Some component slopes contain inflection points.

We first examine two component slopes of the dc differential amplifier studied previously:

$E1 := 1$ $\qquad$ $E2 := -1$ $\qquad$ $R1 := 10$ $\qquad$ $R2 := 100$ $\qquad$ $R3 := 10$

$R4 := 100$ $\qquad$ $W := 50$ $\qquad$ (50 points)

$$G(R1,R2,R3,R4,E1,E2) := \frac{E1 \cdot \left(1 + \frac{R2}{R1}\right)}{1 + \frac{R3}{R4}} - \frac{E2 \cdot R2}{R1}$$

Vary R1 and R2 from 80% to 120%:

$$m1 := 0.8 \qquad m2 := 1.2 \qquad k := 1..W+1$$

$$R1v_k := R1 \cdot \left[\frac{(m2-m1)}{W} \cdot (k-1) + m1\right]$$

$$V1_k := G(R1v_k, R2, R3, R4, E1, E2)$$

$$R2v_k := R2 \cdot \left[\frac{(m2 - m1)}{W} \cdot (k-1) + m1 \right]$$

$$V2_k := G(R1, R2v_k, R3, R4, E1, E2)$$

Plot output versus component value change   (No inflection points):

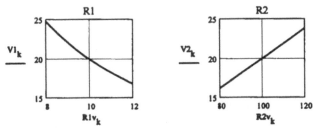

FIGURE 37.   R1 (left) and R2 (right).

Component slope for BPF capacitor C1:

$$K := 10^3 \qquad uF := 10^{-6}$$

$$R1 := 6.34 \cdot K \qquad R2 := 80.6 \qquad R3 := 127 \cdot K$$

$$1 := 0.1 \cdot uF \qquad C2 := C1 \qquad P := 2 \cdot \pi$$

Choose 490 Hz:

$$F1 := 490 \qquad s1 := P \cdot F1 \cdot \sqrt{-1}$$

$$G(R1, R2, R3, C1, C2, s) :=$$

$$\frac{\dfrac{s_1}{R1 \cdot C1}}{s^2 + \dfrac{s}{R3} \cdot \left(\dfrac{1}{C1} + \dfrac{1}{C2} \right) + \dfrac{1}{R3 \cdot C1 \cdot C2} \cdot \left(\dfrac{1}{R1} + \dfrac{1}{R2} \right)}$$

Vary C1 80% to 120%:

$$C1v_k := C1 \cdot \left[\frac{(m2 - m1)}{W} \cdot (k-1) + m1 \right]$$

$$V1_k := |\, G(R1, R2, R3, C1v_k, C2, s1)\,|$$

The inflection point for C1 is very evident in the plot below.

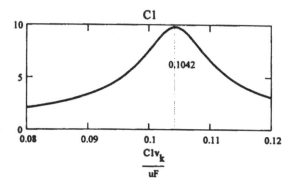

FIGURE 38. C1.

The maximum output occurs not at the nominal value of C1, nor at the 10% value but instead occurs at +4.2%, an intermediate value as implied in Figures 35 and 36. Given enough samples N, the probability is high that a MCA will find this 4.2% point (and for the other components) and display the higher output.

The derivative of C1 with respect to the output at a constant frequency is:

$$C1b := \frac{C2 \cdot \left[R3 \cdot \left(\frac{1}{R1} + \frac{1}{R2} \right) - 1 \right]}{(P \cdot F1 \cdot R3 \cdot C2)^2 + 1} \qquad C1b = 0.104 \circ uF$$

$$\frac{C1b}{C1} - 1 = 4.24\,\%$$

which verifies the inflection point plot above.

Hence taking the derivatives of each component in sequence should give a true EVA. We attempt to do this below:

$$R2b := \frac{R1}{(P \cdot F1)^2 \cdot R1 \cdot R3 \cdot C1b \cdot C2 - 1} \qquad R2b = 80.7$$

$$R3b := \frac{(R1 + R2b)^2 + (P \cdot F1 \cdot R1 \cdot R2b \cdot (C1b + C2))^2}{(P \cdot F1)^2 \cdot R1 \cdot R2b \cdot C1b \cdot C2 \cdot (R1 + R2b)}$$

$$R3b = 127318.9$$

$R1b := 0.98 \cdot R1$ (R1 sensitivity is −1 across the frequency band.)

$$C2b := \frac{(R1 + R2b)^2 + (P \cdot F1 \cdot R1b \cdot R2b \cdot C1b)^2}{(P \cdot F1)^2 \cdot R2b \cdot C1b \cdot \left[R3b \cdot \left(R1b \cdot R2b + R1^2 \right) - R1b^2 \cdot R2b \right]}$$

$C2b = 0.096\,\text{∘uF}$

$Vo := |G(R1, R2, R3, C1, C2, s1)|$ $Vo = 7.612$ Nominal

Put all the maximized components together:

$Ve := |G(R1b, R2b, R3b, C1b, C2b, s1)|$ $Ve = 7.777$ EVA value?

$Va := |G(R1b, R2, R3, C1b, C2, s1)|$ $Va = 10.003$

No, R1b and C1b gives a higher amplitude.

Try taking derivatives using nominal values:

$$R2a := \frac{R1}{(P \cdot F1)^2 \cdot R1 \cdot R3 \cdot C1 \cdot C2 - 1} \qquad R2a = 84.173$$

$$\frac{R2a}{R2} - 1 = 4.43\,\%$$

$$R3a := \frac{(R1+R2)^2 + (P \cdot F1 \cdot R1 \cdot R2 \cdot (C1+C2))^2}{(P \cdot F1)^2 \cdot R1 \cdot R2 \cdot C1 \cdot C2 \cdot (R1+R2)}$$

$$R3a = 132874.36$$

$$C1a := C1b \qquad R1a := R1b$$

$$C2a := \frac{(R1+R2b)^2 + (P \cdot F1 \cdot R1 \cdot R2 \cdot C1)^2}{(P \cdot F1)^2 \cdot R2 \cdot C1 \cdot \left[R3 \cdot \left(R1 \cdot R2 + R1^2\right) - R1^2 \cdot R2\right]}$$

$$C2a = 0.105 \ \text{°uF}$$

$$Vf := |G(R1a, R2a\,R3a, C1a, C2a, s1)| \qquad Vf = 3.827$$

This obviously doesn't work either.

The additional constraints of remaining within the component tolerances have been neglected. We have violated this constraint for R2a at +4.43%.

Hence the assertion remains that a true EVA algorithm for this circuit does not exist. Only an infinite value of N in an MCA will find the component set producing the highest possible output at a given frequency.

A quote from a Spice worst-case analysis manual referring to the .WCASE feature is germane here: "Worst-case analysis is not an optimization process. That is, it does not search for the set of parameter values which result in the worst result. It assumes that the worst case will occur when each parameter (component) has been either pushed to one of its limits or left at its nominal value as indicated by the sensitivity analysis. It will show the true worst-case results when the collating function is monotonic within all tolerance combinations. Otherwise, there is no guarantee."

SALLEN AND KEY BPF

This ac circuit contains several surprises and is given for that reason. The circuit, a Sallen and Key bandpass filter is well known and extensively described in active filter literature. The circuit was designed per Reference 11, using parameters of gain and Q within that recommended for this design. A RSS/EVA analysis is performed first with a by now familiar sequence. These results and then compared to both MCA and FMCA.

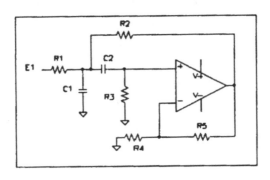

$$K := 103 \qquad uF := 10{-}6 \qquad \text{Gain} = 10; \ Q = 10$$

$$fo := 500 \qquad P := 2 \cdot \pi$$

$$wo := P \cdot fo \qquad C1 := 0.1 \cdot uF \qquad C2 := C1$$

$$R1 := 15.8 \cdot K \qquad R2 := 5.11 \cdot K \qquad R3 := 2.61 \cdot K$$

$$R4 := 3.32 \cdot K \qquad R5 := 13.3 \cdot K$$

$$BF := 400 \qquad LF := 600 \qquad DF := 1 \qquad i := 1..\left(\frac{LF - BF}{DF}\right) + 1$$

$$F_i := BF + DF \cdot (i - 1) \qquad s_i := P \cdot F_i \cdot \sqrt{-1}$$

$$G(R1, R2, R3, R4, R5, C1, C2, s) :=$$

$$\frac{\dfrac{s_i \cdot (R4 + R5)}{R4 \cdot R1 \cdot C1}}{\left(s_i\right)^2 + \dfrac{s_i}{C1} \cdot \left(\dfrac{1}{R1} + \dfrac{2}{R3} - \dfrac{R5}{R2 \cdot R4}\right) + \dfrac{1}{R3 \cdot C1 \cdot C2} \cdot \left(\dfrac{1}{R1} + \dfrac{1}{R2}\right)}$$

$$Vo_i := |G(R1, R2, R3, R4, R5, C1, C2, s)|$$

$$Nc := 7 \qquad dpf := 0.0001 \qquad p := 1..Nc$$

$$Q := dpf \cdot identity(Nc) + 1$$

$$Vr_{i,p} := \left| G\left(\begin{matrix} R1 \cdot Q_{p,1}, R2 \cdot Q_{p,2}, R3 \cdot Q_{p,3}, R4 \cdot Q_{p,4}, R5 \cdot Q_{p,5}, \\ C1 \cdot Q_{p,6}, C2 \cdot Q_{p,7}, s \end{matrix}\right) \right|$$

$$Sen_{i,p} := \left(\frac{Vr_{i,p}}{Vo_i} - 1\right) \cdot \frac{1}{dpf} \qquad Tc := 0.1 \qquad Tr := 0.02$$

$$T := \begin{bmatrix} -Tr & -Tr & -Tr & -Tr & -Tr & -Tc & -Tc \\ Tr & Tr & Tr & Tr & Tr & Tc & Tc \end{bmatrix}$$

$$L_{i,p} := if(Sen_{i,p} > 0, 1 + T_{1,p}, 1 + T_{2,p})$$

$$H_{i,p} := if(Sen_{i,p} > 0, 1 + T_{2,p}, 1 + T_{1,p})$$

RSS Low:

$$Vrssl_i := Vo_i \cdot \left[1 - \sqrt{\sum_p \left[Sen_{i,p} \cdot \left(L_{i,p} - 1\right)\right]^2}\right]$$

RSS High:

$$Vrss2_i := Vo_i \cdot \left[1 + \sqrt{\sum_p \left[Sen_{i,p} \cdot \left(H_{i,p} - 1 \right) \right]^2} \right]$$

EVL:

$$VL_i := \left| G\left(\begin{matrix} R1 \cdot L_{i,1}, R2 \cdot L_{i,2}, R3 \cdot L_{i,3}, R4 \cdot L_{i,4}, R5 \cdot L_{i,5}, \\ C1 \cdot L_{i,6}, C2 \cdot L_{i,7}, s \end{matrix} \right) \right|$$

EVH:

$$VH_i := \left| G\left(\begin{matrix} R1 \cdot H_{i,1}, R2 \cdot H_{i,2}, R3 \cdot H_{i,3}, R4 \cdot H_{i,4}, R5 \cdot H_{i,5}, \\ C1 \cdot H_{i,6}, C2 \cdot H_{i,7}, s \end{matrix} \right) \right|$$

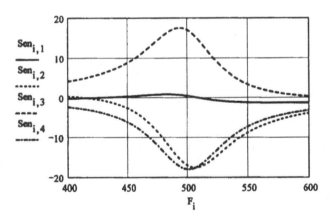

FIGURE 39. Sensitivities — R1 through R4.

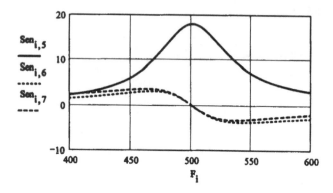

FIGURE 40.   Sensitivities — R5, C1, and C2.

Note sensitivities > 10 and that most do not pass through zero at
fo = 500 Hz.

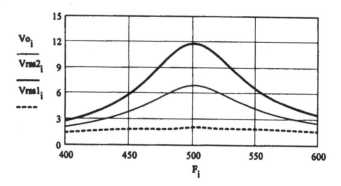

FIGURE 41.   Vo + 3s and Vo.

High ST products again cause misleading RSS analysis. Other than nominal Vo, this plot should be ignored.

For EVA:

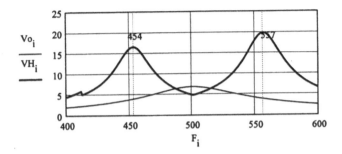

FIGURE 42. EVA high and Vo.

Is this plot anywhere near correct? The answer is **no**.

The only redeeming feature of the EVA plot of Figure 8 was that the extreme center frequencies were correct. Not even these are correct for this design as shown below.

EVH center frequency calculations:

$$D0 := \frac{1}{R3 \cdot C1 \cdot C2} \cdot \left(\frac{1}{R1} + \frac{1}{R2} \right) \qquad fo := \frac{\sqrt{D0}}{P}$$

$$fo = 501.3 \qquad \text{Nominal}$$

$$foh := \frac{\sqrt{\dfrac{1}{R3 \cdot 1.02 \cdot C1 \cdot 0.9 \cdot C2 \cdot 0.9} \cdot \left(\dfrac{1}{R1 \cdot 0.98} + \dfrac{1}{R2 \cdot 0.98} \right)}}{P}$$

$$foh = 557.2 \qquad Sen_3 > 0$$

Figure 42 above shows the high center frequency at 557 Hz, that calculated above with Sen_3 for R3 > 0, while the remaining components have sensitivities < 0. If we were to recalculate the extreme center frequency using Sen_3 < 0, as in an FMCA, all components would then have the low tolerance assigned as in the following:

$$fah := \frac{\sqrt{\dfrac{1}{R3 \cdot 0.98 \cdot C1 \cdot 0.9 \cdot C2 \cdot 0.9}\left(\dfrac{1}{R1 \cdot 0.98} + \dfrac{1}{R2 \cdot 0.98}\right)}}{P}$$

$fah = 568.4$ All Sen > 0

We find this combination produces a higher center frequency (568.4) than the so-called EVA high. This is exactly what happens when an FMCA is performed. One final comment about this circuit before proceeding to an FMCA. Notice that the design center frequency amplitude was 10V. The precise design values for the components were as follows (see Reference 11.):

$R1 := 15915.49$ $R2 := 5064.6$ $R3 := 2637.2$

$R4 := 3296.5$ $R5 := 13185.99$

With the exception of R1, the four resistor sensitivities are > 10 at F = 500 Hz. Hence when converting to the standard 1% values, the desired amplitude was not achieved. In fact it was quite far off:

$$F_{101} = 500, \quad Vo_{101} = 6.941$$

This is another drawback of this design, though not necessarily of this topology.

We saw that in the case of the multiple feedback BPF, that MCA saved the day for us and gave a reliable estimate of the worst-case performance of the circuit. However, this is not the case with this filter. There is one more surprise in store for us.

SALLEN AND KEY BPF — FMCA AND EVA COMBINED

See page 98 for G-function setup.

$Vo_i := |G(R1, R2, R3, R4, R5, C1, C2, s)|$

$Nc := 7 \quad dpf := 0.0001 \quad p := 1..Nc \quad Q := dpf \cdot identity(Nc) + 1$

$$Vr_{i,p} := \left| G\left(\begin{matrix} R1 \cdot Q_{p,1}, R2 \cdot Q_{p,2}, R3 \cdot Q_{p,3}, R4 \cdot Q_{p,4}, R5 \cdot Q_{p,5}, \\ C1 \cdot Q_{p,6}, C2 \cdot Q_{p,7}, s \end{matrix} \right) \right|$$

$$Sen_{i,p} := \left(\frac{Vr_{i,p}}{Vo_i} - 1 \right) \cdot \frac{1}{dpf} \quad Tc := 0.1 \quad Tr := 0.02$$

$$T := \begin{bmatrix} -Tr & -Tr & -Tr & -Tr & -Tr & -Tc & -Tc \\ Tr & Tr & Tr & Tr & Tr & Tc & Tc \end{bmatrix}$$

$L_{i,p} := if(Sen_{i,p} > 0, 1 + T_{1,p}, 1 + T_{2,p})$

$H_{i,p} := if(Sen_{i,p} > 0, 1 + T_{2,p}, 1 + T_{1,p})$

EVA:

$$VL_i := \left| G\left(\begin{matrix} R1 \cdot L_{i,1}, R2 \cdot L_{i,2}, R3 \cdot L_{i,3}, R4 \cdot L_{i,4}, R5 \cdot L_{i,5}, \\ C1 \cdot L_{i,6}, C2 \cdot L_{i,7}, s \end{matrix} \right) \right|$$

$$VH_i := \left| G\left(\begin{matrix} R1 \cdot H_{i,1}, R2 \cdot H_{i,2}, R3 \cdot H_{i,3}, R4 \cdot H_{i,4}, R5 \cdot H_{i,5}, \\ C1 \cdot H_{i,6}, C2 \cdot H_{i,7}, s \end{matrix} \right) \right|$$

FMCA:

$N := 2^{Nc} \quad k := 1..N \quad w := 1..Nc \quad Re_{k,w} := k \quad N = 128$

$$Re_{k,w+1} := \text{floor}\left(\frac{Re_{k,w}}{2}\right) \qquad Dr_{k,w} := Re_{k,w} - 2 \cdot Re_{k,w} + 1$$

$$Tn_{k,w} := \text{if}\left(Dr_{k,w} = 0, 1 + T_{1,w}, 1 + T_{2,w}\right)$$

$$Vm_{k,i} := \left| G\left(\begin{array}{c} R1 \cdot Rn_{k,1}, R2 \cdot Tn_{k,2}, R3 \cdot Tn_{k,3}, R4 \cdot Tn_{k,4}, R5 \cdot Tn_{k,5}, \\ C1 \cdot Tn_{k,6}, C2 \cdot Tn_{k,7}, s \end{array}\right) \right|$$

$$Vmax_i := \max\left(Vm^{<i>}\right) \qquad Vpk := \max(Vmax) \qquad Vpk = 28.222$$

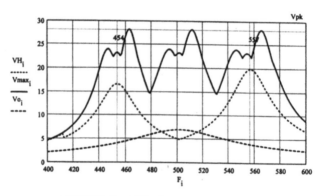

FIGURE 43. EVA (dot) and FMCA (solid).

Note the extreme disparity between the EVA and FMCA plots. The reader would do well to keep these ac examples in mind. If he or she is a member of a contract negotiating team with a potential customer, and that customer stipulates EVA for all ac circuits, it might be productive to demonstrate the anomalous EVA performance of some ac circuits, possibly using the examples herein. The goal is to convince the customer that MCA would better serve the intent of the worst-case analysis requirements.

MCA – SALLEN AND KEY BPF
(Using uniform distribution)

See page 98 for G-function and tolerances.

$$Vo_i := |F(R1, R2, R3, R4, R5, C1, C2, s)|$$

$$N := 2000 \qquad Nc := 7 \qquad k := 1..N \qquad w := 1..Nc$$

$$Tn_{k,w} := (T_{2,w} - T_{1,w}) \cdot rnd(1) + T_{1,w} + 1$$

$$Vm_{k,i} := \left| G\left(\begin{array}{l} R1 \cdot Rn_{k,1}, R2 \cdot Tn_{k,2}, R3 \cdot Tn_{k,3}, R4 \cdot Tn_{k,4}, R5 \cdot Tn_{k,5}, \\ C1 \cdot Tn_{k,6}, C2 \cdot Tn_{k,7}, s \end{array}\right) \right|$$

$$Vmax_i := max(Vm^{<i>})$$

The somewhat unexpected output waveshapes are shown below.

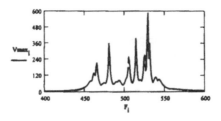

FIGURE 44.   Run #1   (N = 2000).

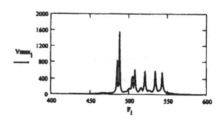

FIGURE 45.   Run #2   (N = 2000).

WHY MCA SPIKES OCCUR

The magnitude of any BPF transfer function is:

$$Mag = \frac{N1 \cdot \omega}{\sqrt{\left(D0 - \omega^2\right)^2 + \left(D1 \cdot \omega\right)^2}}$$

At the center frequency

$$D0 = \omega^2$$

The magnitude is then:

$$Mag = \frac{N1}{D1}$$

The nominal resistor values are:

R1 = 15.8 ∘K   R2 = 5.11 ∘K   R3 = 2.61 ∘K

R4 = 3.32 ∘K   R5 = 13.3 ∘K

$$N1 := \frac{R4 + R5}{R1 \cdot R4 \cdot C1} \qquad N1 = 3168.37$$

$$D1 := \frac{1}{C1} \cdot \left(\frac{1}{R1} + \frac{2}{R3} - \frac{R5}{R2 \cdot R4} \right) \qquad D1 = 456.169$$

$$Mag := \frac{N1}{D1} \qquad Mag = 6.946$$

If D1 becomes very small, then the center frequency magnitude becomes very large. Note the minus sign in D1. If we can find toleranced values of R2, R4, and R5 such that

$$\frac{R5}{R2 \cdot R4} = \frac{1}{R1} + \frac{2}{R3}$$

then D1 = 0 and the magnitude is infinite. It turns out this is not very hard to do.

$$R2a := 0.982 \cdot R2 \qquad R4a := 0.9816 \cdot R4 \qquad R5a := 1.02 \cdot R5$$

$$\text{(All at or within the 2\% tolerance)}$$

$$D1a := \frac{1}{C1} \cdot \left(\frac{1}{R1} + \frac{2}{R3} - \frac{R5a}{R2a \cdot R4a} \right) \qquad D1a = 0.166$$

$$N1a := \frac{R4a + R5a}{R1 \cdot R4a \cdot C1} \qquad N1a = 3267.56$$

$$Mag := \frac{N1a}{D1a} \qquad mag = 19688.428 \qquad 20 \cdot \log(Mag) = 85.88 \quad dBV$$

Hence the culprit is the minus sign in D1. *Caveat:* Beware of transfer functions that have negative terms in the coefficients. Note that the cause of the spikes would be very difficult to determine using matrix analysis. Hence low-order transfer functions should be utilized if available.

ESTIMATING MANUFACTURING YIELD (MY)

MCA provides one of the best ways to estimate the yield in a manufacturing run. Using MCA may avoid the necessity of an expensive manufacturing pilot run of a small number of circuit boards. Use a high number of samples N given time and computer memory limitations. The statistics given in the higher samples will be more accurate and reliable. See "Confidence Intervals" in Part 3.

We first do a manufacturing yield analysis on the dc differential amplifier analyzed in Part 1. Mathcad programming is used to find the MY using three different resistor tolerances. Hence this could determine the least expensive components to meet a specified MY.

Note that 20,000 samples are used with a uniform distribution, which is more conservative than a normal distribution. The nominal output of this circuit is Vo = 20V and the hypothetical test limits are

$$LL = 19.5V \text{ and } UL = 20.5$$

$$E1 := 1 \quad E2 := -1 \quad R1 := 10 \quad R2 := 100 \quad R3 := 10 \quad R4 := 100$$

$$G(R1, R2, R3, R4, E1, E2) := \frac{E1 \cdot \left(1 + \dfrac{R2}{R1}\right)}{1 + \dfrac{R3}{R4}} - \frac{E2 \cdot R2}{R1}$$

Resistor tolerances: 1, 2, and 5%

$$Re := (0.01 \quad 0.02 \quad 0.05)^{\mathsf{T}} \qquad Te := 0$$

Input voltage tolerance is zero so that the only failures occur due to the resistor tolerances

$$N := 20000 \qquad k := 1..N \qquad Nc := 6 \qquad w := 1..Nc$$

$$LL := 19.5 \qquad UL := 20.5$$

Three nested loops are used. The outer loop (counter a) performs the MCA using each of the three resistor tolerances. The middle loop (k) counts through the N samples, while the final inner loop (w) assigns the random tolerances to each of the Nc components.

$$yld := \begin{vmatrix} \text{for } a \in 1..3 \\ \quad fail \leftarrow 0 \\ \quad Tr \leftarrow Re_a \\ \quad Trm_a \leftarrow Tr \\ \quad T \leftarrow \begin{bmatrix} -Tr & -Tr & -Tr & -Tr & -Te & -Te \\ Tr & Tr & Tr & Tr & Te & Te \end{bmatrix} \\ \quad \begin{vmatrix} \text{for } k \in 1..N \\ \quad \begin{vmatrix} \text{for } w \in 1..Nc \\ \quad Tn_{k,w} \leftarrow \left(T_{2,w} - T_{1,w}\right)\cdot rnd(1) + T_{1,w} + 1 \\ \quad Vm_k \leftarrow G\left(R1\cdot Tn_{k,1}, R2\cdot Tn_{k,2}, R3\cdot Tn_{k,3}, R4\cdot Tn_{k,4}, E1\cdot Tn_{k,5}, E2\cdot Tn_{k,6}\right) \\ \quad fail \leftarrow fail + 1 \text{ if } \left(Vm_k < LL\right) + \left(Vm_k > UL\right) \end{vmatrix} \\ \quad fcnt_a \leftarrow fail \end{vmatrix} \\ \begin{bmatrix} Trm \\ fcnt \\ 1 - \dfrac{fcnt}{N} \end{bmatrix} \end{vmatrix}$$

$$yld = \begin{bmatrix} \{3,1\} \\ \{3,1\} \\ \{3,1\} \end{bmatrix}$$

$$pcy := augment\left(yld_1, augment\left(yld_2, yld_3\right)\right)$$

$$pcy^T = \begin{bmatrix} 0.01 & 0.02 & 0.05 \\ 0 & 2511 & 10915 \\ 1 & 0.874 & 0.454 \end{bmatrix}$$

1st row : Resistor tolerances in decimal %

2nd row : Fail count at each tolerance

3rd row : Yield at each tolerance

Hence if the specified MY is 95%, 1% resistors must be used. We next perform an MY analysis on the multiple-feedback BPF.

BPF MANUFACTURING YIELD ANALYSIS

In this analysis, a normal distribution is used. The reader is encouraged to rerun this analysis using a uniform distribution to show the decrease in yield. The component tolerances can be manually varied until the specified MY is met.

$K := 10^3$ $uF := 10^{-6}$ $R1 := 6.34 \cdot K$ $R2 := 80.6$

$R3 := 127 \cdot K$ $C := 0.1 \cdot uF$ $C2 := C1$

$BF := 450$ $LF := 550$ $DF := 10..$ $lit := \dfrac{LF - BF}{DF} + 1$

$i := 1..lit$ $F_i := BF + DF \cdot (i - 1)$

$s_i := 2 \cdot \pi \cdot F_i \cdot \sqrt{-1}$ $N := 2000$ $Nc := 5$ $Tr := 0.01$

$Tc := 0.05$ $T := \begin{bmatrix} -Tr & -Tr & -Tr & -Tc & -Tc \\ Tr & Tr & Tr & Tc & Tc \end{bmatrix}$

$k := 1..N$ $w := 1..Nc$ $z_w := rnorm(N, 0, 1)$

$$Tn_{k,w} := \frac{T_{2,w} - T_{1,w}}{6} \cdot \left[(z_w)_k + 3 \right] + T_{1,w} + 1$$

$$G(R1, R2, R3, C1, C2, s) :=$$

$$\frac{\dfrac{s_i}{R1 \cdot C1}}{(s_i)^2 + \dfrac{s_i}{R3} \cdot \left(\dfrac{1}{C1} + \dfrac{1}{C2} \right) + \dfrac{1}{R3 \cdot C1 \cdot C2} \cdot \left(\dfrac{1}{R1} + \dfrac{1}{R2} \right)}$$

$$Vm_{k,i} := \left| G \begin{pmatrix} R1 \cdot Tn_{k,1}, R2 \cdot Tn_{k,2}, R3 \cdot Tn_{k,3}, \\ C1 \cdot Tn_{k,4}, C2 \cdot Tn_{k,5}, s \end{pmatrix} \right|$$

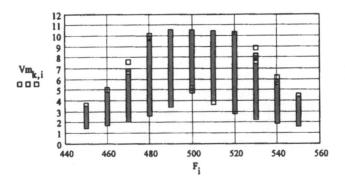

FIGURE 46. Min/max.

This plot shows what might be described as a top view of a series of 3-D histograms at every 10 Hz from 450 Hz to 550 Hz.

Limit matrix:

$$
LM := \begin{bmatrix}
450 & 460 & 470 & 480 & 490 & 500 & 510 & 520 & 530 & 540 & \mathbf{550} \\
3 & 5 & 6 & 9 & 10.5 & 11 & 10.5 & 9 & 6 & 5 & 4 \\
1 & 2 & 2.5 & 3 & 4 & 5.5 & 4 & 3 & 2.5 & 2 & 1
\end{bmatrix}^T
$$

1st row: Freq (Hz)
2nd row: Upper Limit
3rd row: Lower Limit

The limit matrix is created to show what the upper and lower limits of the BPF are at every 10 Hz. These number of course would originate from the functional test requirements of the BPF.

$$
nc := \text{rows}(LM) \qquad nc = 11 \qquad a := 1 .. nc \qquad n_a := \frac{LM_{a,1} - BF}{DF} + 1
$$

$$
yld := \begin{array}{|l}
\text{for } u \in 1 .. nc \\
\quad \begin{array}{|l}
\text{fail} \leftarrow 0 \\
\text{for } k \in 1 .. N \\
\quad \text{fail} \leftarrow \text{fail} + 1 \quad \text{if } \left(Vm_{k,n_u} < LM_{u,3} \right) + \left(Vm_{k,n_u} > LM_{u,2} \right) \\
\text{fcnt}_u \leftarrow \text{fail}
\end{array} \\
\begin{bmatrix} LM \\ \text{fcnt} \end{bmatrix}
\end{array}
$$

$$
yld := \text{augment}(yld_1, yld_2) \qquad N = 2000
$$

$$yld^T = \begin{bmatrix} 450 & 460 & 470 & 480 & 490 & 500 & 510 & 520 & 530 & 540 & 550 \\ 3 & 5 & 6 & 9 & 10.5 & 11 & 10.5 & 9 & 6 & 5 & 4 \\ 1 & 2 & 2.5 & 3 & 4 & 5.5 & 4 & 3 & 2.5 & 2 & 1 \\ 29 & 4 & 20 & 17 & 4 & 5 & 1 & 36 & 54 & 8 & 4 \end{bmatrix}$$

1st row: Freq (Hz)
2nd row: UL
3rd row: LL
4th row: No. failed

$$fsum := \sum_{p=1}^{nc} yld_{p,4} \qquad fsum = 182$$

$$pcyld := 1 - \frac{fsum}{N} \qquad pcyld = 90.9 \cdot \%$$

3 Advanced Topics

Calculating the normalized sensitivities requires dividing by the nominal output voltage Vo. It is possible that in some circuits the nominal output could be zero. Hence we must work around this to perform an EVA/RSS.

DIFFERENTIAL AMPLIFIER, VO = 0

Ampplication: rejection of common mode noise (E1 = E2) (see Reference 13).

$E1 := 1$ $E2 := 1$ $R1 := 10$ $R2 := 100$ $R3 := 10$

$R4 := 100$ $mV := 10^{-3}$

$$G(R1, R2, R3, R4, E1, E2) := \frac{E1 \cdot \left(1 + \dfrac{R2}{R1}\right)}{1 + \dfrac{R3}{R4}} - \frac{E2 \cdot R2}{R1}$$

$Vo := G(R1, R2, R3, R4, E1, E2)$

$Vo = 0$ $Nc := 4$ $p := 1 .. Nc$ $dpf := 0.0001$

$Q := dpf \cdot identity(Nc) + 1$

$Vr_p := G(R1 \cdot Q_{p,1}, R2 \cdot Q_{p,2}, R3 \cdot Q_{p,3}, R4 \cdot Q_{p,4}, E1, E2)$

$Tr := 0.001$

Precision resistors; input voltage tolerance zero to determine CMR of circuit – ideal opamp.

$$T := \begin{bmatrix} -Tr & -Tr & -Tr & -Tr \\ Tr & Tr & Tr & Tr \end{bmatrix} \qquad k := 1..2$$

Note use of Vr instead of Sen in M array:

$$M_{k,p} := if(k = 1, if(Vr_p > 0, 1 + T_{1,p}, 1 + T_{2,p}),$$
$$if(Vr_p > 0, 1 + T_{2,p}, 1 + T_{1,p}))$$

$$Vev_k := G(R1 \cdot M_{k,1}, R2 \cdot M_{k,2}, R3 \cdot M_{k,3}, R4 \cdot M_{k,4}, E1, E2)$$

$$Vrss_k := \frac{(-1)^k}{dpf} \cdot \sqrt{\sum_p \left[Vr_p \cdot \left(M_{k,p} - 1 \right) \right]^2}$$

See Eqn (2) in Part 1.

$$Vev = \begin{bmatrix} -3.643 \\ 3.63 \end{bmatrix} \circ mV \qquad Vrss = \begin{bmatrix} -1.818 \\ 1.818 \end{bmatrix} \circ mV$$

Rejection in dBV:

$$20 \cdot \log(Vev_2) = -48.8 \qquad 20 \cdot \log(Vrss_2) = -54.81$$

EVA/RSS OF OPAMP OFFSETS

Zero initial offsets

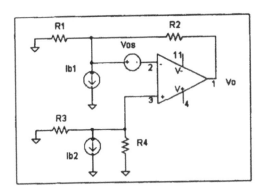

$K := 10^3$ $nA := 10^{-9}$ $mV := 10_{-3}$

$R1 := 10 \cdot K$ $R2 := 200 \cdot K$ $R3 := 10 \cdot K$ $R4 := 200 \cdot K$

Nominal offsets:

$Vos := 0$ $Ib1 := 0$ $Ib2 := 0$ $Ib1 = Ib + \dfrac{Ios}{2}$ $Ib2 = Ib - \dfrac{Ios}{2}$

This is a typical configuration for determining offset errors of an opamp. The resistors could be Thevenin equivalents. Bias current (Ib) and offset current (Ios) are extracted from the opamp data book.

$$G(R1,R2,R3,R4,Vos,Ib1,Ib2) :=$$

$$\left(Vos + \frac{Ib1 \cdot R1 \cdot R2}{R1 + R2} - \frac{Ib2 \cdot R3 \cdot R4}{R3 + R4} \right) \cdot \left(1 + \frac{R2}{R1} \right)$$

$Vo := G(R1, R2, R3, R4, Vos, Ib1, Ib2)$ $Vo = 0 \circ mV$

$$Nc := 7 \qquad p := 1 .. Nc \qquad dpf := 0.0001$$

$$Q := dpf \cdot identity(Nc) + 1 \qquad k := 1 .. 2$$

$$Vr_p := G(R1 \cdot Q_{p,1}, R2 \cdot Q_{p,2}, R3 \cdot Q_{p,3}, R4 \cdot Q_{p,4}, Q_{p,5} - 1,$$
$$Q_{p,6} - 1, Q_{p,7} - 1) \qquad \text{(Note difference)}$$

$$Vr^T = [0 \ 0 \ 0 \ 0 \ 0.002 \ 20 \ -20]$$

Vr_1 thru Vr_4 are zero because of zero input.

Set tolerances for resistors and offsets:

$$Tr := 0.02 \qquad\qquad T5a := 8 \cdot mV \qquad dT5b := 10 \cdot mV$$

$$T6 := 8.8 \cdot nA \qquad\qquad T7 := 3 \cdot nA$$

$$T := \begin{bmatrix} -Tr & -Tr & -Tr & -Tr & -T5a & -T6 & -T7 \\ Tr & Tr & Tr & Tr & T5b & T6 & T7 \end{bmatrix}$$

Note use of Vr instead of Sen in M array:

$$M_{k,p} := if(k = 1, if(Vr_p < 0, 1 + T_{2,p}, 1 + T_{1,p}),$$
$$if(Vr_p \geq 0, 1 + T_{2,p}, 1 + T_{1,p}))$$

$$Vev_k := G(R1 \cdot M_{k,1}, R2 \cdot M_{k,2}, R3 \cdot M_{k,3}, R4 \cdot M_{k,4},$$
$$M_{k,5} - 1, M_{k,6} - 1, M_{k,7} - 1) \qquad \text{(Note difference)}$$

$$Vrss_k := \frac{(-1)^k}{dpf} \sqrt{\sum_p \left[Vr_p \cdot \left(M_{k,p} - 1 \right) \right]^2} \qquad \text{Eqn (2) from Part 1.}$$

$$Vrss = \begin{bmatrix} -0.168 \\ 0.210 \end{bmatrix} \qquad Vev = \begin{bmatrix} -0.170 \\ 0.212 \end{bmatrix}$$

Non-zero initial offsets can be accounted for with assymetric offsets, as in T5a and T5b above. For example, if the initial Vos = 1mV, and the lower and upper limits were –6mV and +7mV, then T5a = –5mV and T5b = 8mV.

The Vrss numbers will be corroborated by an MCA analysis.

MCA OF OPAMP OFFSETS

See previous page for the schematic and G-function setup.

$$Vo := G(R1, R2, R3, R4, Vos, Ib1, Ib2) \qquad Vo = 0 \circ mV \qquad Nc := 7$$

$$N := 20000 \qquad k := 1 .. N \qquad w := 1 .. Nc \qquad z_w := \text{rnorm}(N, 0, 1)$$

$$Tr := 0.02 \qquad T5a := 8 \cdot mV \qquad T5b := 10 \cdot mV \qquad T6 := 8.8 \cdot nA$$

$$T7 := 3 \cdot nA$$

$$T := \begin{bmatrix} -Tr & -Tr & -Tr & -Tr & -T5a & -T6 & -T7 \\ Tr & Tr & Tr & Tr & T5b & T6 & T7 \end{bmatrix}$$

$$Tn_{k,w} := \frac{T_{2,w} - T_{1,w}}{6} \left\{ \left(z_w\right)_k + 3 \right\} + T_{1,w} + 1$$

$$Vm_k := G(R1 \cdot Tn_{k,1}, R2 \cdot Tn_{k,2}, R3 \cdot Tn_{k,3}, R4 \cdot Tn_{k,4}, \\ Tn_{k,5} - 1, Tn_{k,6} - 1, Tn_{k,7} - 1)$$

$$Vavg := \text{mean}(Vm) \qquad Vs := \text{stdev}(Vm)$$

$$VH := \text{max}(Vm) \qquad VL := \text{min}(Vm)$$

$$nb := 40 \qquad q := 1 .. nb + 1 \qquad nh := 1 .. nb \qquad intv := \frac{VH - VL}{nb}$$

$$bin_q := VL + intv \cdot (q - 1)$$

$pd := \text{hist} (\text{bin}, Vn)$ $E(Vm) := N \cdot \text{intv} \cdot \text{dnorm} (Vm, Vavg, Vs)$
$u := 1 .. 2$

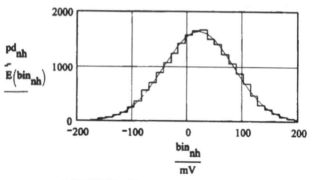

FIGURE 47. Histogram — opamp offsets.

$Vavg = 20.89 \circ mV$ $Vs = 62.87 \circ mV$

$Vmca_u := Vavg + (-1)^u \cdot 3 \cdot Vs$

$$Vmca = \begin{bmatrix} -0.168 \\ 0.210 \end{bmatrix}$$

$$Vrss := \begin{bmatrix} -0.168 \\ 0.210 \end{bmatrix}$$ (From previous page)

$$\overrightarrow{\frac{Vrss}{Vmca}} - 1 = \begin{bmatrix} 0.163 \\ 0.233 \end{bmatrix} \circ \%$$ (Error less than 1%)

If the input was uniform instead of normal:

$$Vrsu := \sqrt{3} \cdot Vmca - Vavg \qquad Vrsu = \begin{bmatrix} -0.311 \\ 0.342 \end{bmatrix}$$

COMPARING NORMAL AND UNIFORM DISTRIBUTIONS

$$x := -5, -4.95 .. 5 \qquad f(x) := dnorm(x, 0, 1)$$

$$g(x) := dnorm\left(x, 0, \sqrt{3}\right) \qquad h(x) := \frac{1}{6} \cdot (\Phi(x+3) - \Phi(x-3))$$

$$c := \sqrt{2 \cdot \pi} \qquad \text{(Normalizing constant)}$$

Peak or maximum values of distributions: normal distribution (solid line in plot) is

$$\frac{1}{c} = 0.399$$

Normal distribution (dotted line) is

$$\frac{1}{c \cdot \sqrt{3}} = 0.23$$

Uniform distribution is

$$\frac{1}{6} = 0.167$$

(Peak amplitudes for the three distributions give an area of 1 for 100% probability.)

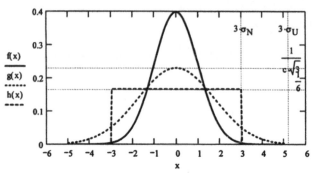

FIGURE 48. Normal and uniform distributions.

Definitions:

$$\Delta Vmca = 3 \cdot \sigma \qquad \Delta Vrss = 3 \cdot \sigma_N \qquad \Delta Vrsu = 3 \cdot \sigma_U$$

Uniform:

$$a := -3 \qquad b := 3 \qquad \sigma_U := \frac{b-a}{\sqrt{12}} \qquad \sigma_U = 1.732$$

For a *normal input* and *normal output*

$$3 \cdot \sigma_N = 3$$

For a *uniform input* and *normal output*

$$3 \cdot \sigma_U = 5.196$$

For a *uniform input* and *uniform output*

$$\frac{3 \cdot \sigma_U}{\sqrt{3}} = 3$$

MCA OF OPAMP OFFSETS (Cont')

Equations same as on page 119.
 Uniform distribution input

$$Tn_{k,w} := (T_{2,w} - T_{1,w}) \cdot rnd(1) + T_{1,w} + 1 \qquad N = 20000$$

$$Vm_k := G(R1 \cdot Tn_{k,1}, R2 \cdot Tn_{k,2}, R3 \cdot Tn_{k,3}, R4 \cdot Tn_{k,4},$$
$$Tn_{k,5} - 1, Tn_{k,6} - 1, Tn_{k,7} - 1)$$

$$Vavg := mean(Vm) \qquad Vs := stdev(Vm)$$

$$VH := max(Vm) \qquad VL := min(Vm)$$

$$nb := 40 \qquad q := 1 .. nb + 1 \qquad nh := 1 .. nb \qquad intv := \frac{VH - VL}{nb}$$

$$bin_q := VL + intv \cdot (q - 1)$$

$$pd := hist(bin, Vm) \qquad u := 1 .. 2$$

From page 118:

$$Vrss := \begin{bmatrix} -0.168 \\ 0.210 \end{bmatrix}$$

Shown below is a *uniform output* for a *uniform input*. Note that the edges of the distribution are the same as that for a normal input/output, or Vrss. If we calculate Vmca we get:

$$Vmca_u := Vavg + (-1)^u \cdot 3 \cdot Vs \qquad Vmca = \begin{bmatrix} -0.305 \\ 0.347 \end{bmatrix}$$

From page 120:

$$Vrsu = \begin{bmatrix} -0.311 \\ 0.342 \end{bmatrix}$$

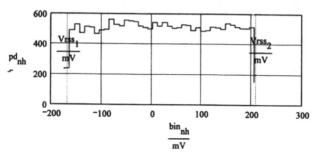

FIGURE 49. Uniform input/output.

The predicted numbers are not as close, but these numbers are for a uniform input/normal output. Reference page 122: For a uniform input and output, the 3σ points are:

$$\frac{3 \cdot \sigma_U}{\sqrt{3}} = \frac{3 \cdot \sqrt{3 \cdot \sigma_N}}{\sqrt{3}} = 3 \cdot \sigma_N$$

TOLERANCE ANALYSIS OF LM158 OPAMP STABILITY/PHASE MARGIN*

To demonstrate that these procedures are not restricted to active filters, but apply to any ac circuit, the stability of an opamp model embedded in a circuit is analyzed as follows:

---

* For an excellent text on automatic control and stability, see *Modern Control Systems*, 7th edition, Dorf and Bishop, Addison-Wesley, 1995.

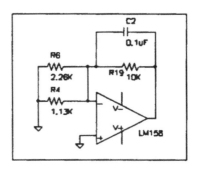

$K := 10^3 \qquad M := 10^6 \qquad uF := 10^{-6} \qquad pF := 10^{-12}$

$Rx := 101.2433 \cdot M \qquad Cx := 80 \cdot pF \qquad \text{(Internal to IC)}$

(Marginal stability. Cx = 200pF improves stability.)

$R4 := 1.13 \cdot K \qquad R6 := 2.26 \cdot K \qquad R19 := 10 \cdot K$

$C2 := 0.1 \cdot uF \qquad \omega1 := \dfrac{1}{Rx \cdot Cx} \qquad p := 2 \cdot \pi$

$\omega1 = 123.5 \qquad \omega2 := P \cdot 1.2 \cdot M \qquad \omega3 := P \cdot 2 \cdot M$

$\omega1$, $\omega2$, and $\omega3$ are the opamp model internal poles.

$BF := 0 \qquad ND := 7 \qquad PD := 10 \qquad i := 1 .. ND \cdot PD + 1$

$L_i := BF + \dfrac{i-1}{PD}$

$F_i := 10^{L} \qquad s_i := \sqrt{-1} \cdot P \cdot F_i \qquad db(x) := 20 \cdot \log(|x|)$

$Ao := 10^5$

Stability can be measured in terms of phase margin. A rule of thumb is that the phase margin should not be less than 45 degrees. See Chapter 6 in the Dorf and Bishop text.

$G(R4, R6, R19, C2, Ao, w1, s) :=$

$$A_i \leftarrow \frac{Ao}{\left(1 + \frac{s_i}{\omega 1}\right) \cdot \left(1 + \frac{s_i}{\omega 2}\right) \cdot \left(1 + \frac{s_i}{\omega 3}\right)}$$ Opamp frequency response

$$B_i \leftarrow \frac{s_i + \frac{1}{C2} \cdot \left(\frac{1}{R4} + \frac{1}{R6} + \frac{1}{R19}\right)}{s_i + \frac{1}{R19 \cdot C2}}$$ Feedback (inverse beta)

$$AB_i \leftarrow \frac{A_i}{B_i}$$ Loop gain

$$\phi_i \leftarrow \arg(AB_i)$$ Loop gain phase

if $\phi_i > 0$

$\quad \left| \begin{array}{l} \phi_i \leftarrow \phi_i - 2 \cdot \pi \\ \phi_i \text{ otherwise} \end{array} \right.$

$$\begin{bmatrix} A_i \\ B_i \\ AB_i \\ \phi_i \end{bmatrix}$$

$$A_i := G(R4, R6, R19, C2, Ao, \omega 1, s)_1$$

$$B_i := G(R4, R6, R19, C2, Ao, \omega 1, s)_2$$

$$AB_i := G(R4, R6, R19, C2, Ao, \omega1, s)_3$$

$$\phi_i := G(R4, R6, R19, C2, Ao, \omega1, s)_4$$

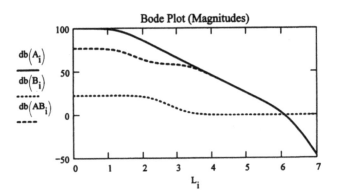

FIGURE 50. Bode plot (magnitudes).

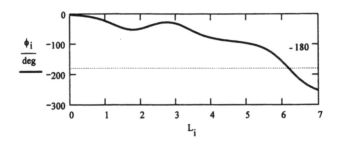

FIGURE 51. Phase (deg).

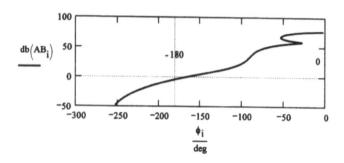

FIGURE 52. Magnitude vs. phase.

R4   R6   R19   C2   Ao   ω1   Column order of tolerances

$$T := \begin{bmatrix} -0.02 & -0.02 & -0.02 & -0.1 & -0.2 & -0.1 \\ 0.02 & 0.02 & 0.02 & 0.1 & 0.2 & 0.1 \end{bmatrix}$$

$Nc := 6 \qquad p := 1 .. Nc \qquad dpf := 0.0001 \qquad Q := dpf \cdot identity(Nc) + 1$

$$ABr_{i,p} := G(R4 \cdot Q_{p,1}, R6 \cdot Q_{p,2}, R19 \cdot Q_{p,3}, C2 \cdot Q_{p,4},$$
$$Ao \cdot Q_{p,5}, \omega1 \cdot Q_{p,6}, s)_3$$

$$Sen_{i,p} := \left(\frac{|ABr_{i,p}|}{|AB_i|} - 1 \right) \cdot \frac{1}{dpf}$$

EVA:

$$L_{i,p'} := if(Sen_{i,p} > 0, 1 + T_{1,p}, 1 + T_{2,p})$$

$$H_{i,p'} := if(Sen_{i,p} > 0, 1 + T_{2,p}, 1 + T_{1,p})$$

$$ABL_i := G(R4 \cdot L_{i,1}, R6 \cdot L_{i,2}, R19 \cdot L_{i,3}, C2 \cdot L_{i,4},$$
$$Ao \cdot L_{i,5}, \omega1 \cdot L_{i,6}, s)_3$$

$$ABH_i := G(R4 \cdot H_{i,1}, R6 \cdot H_{i,2}, R19 \cdot H_{i,3}, C2 \cdot H_{i,4},$$
$$Ao \cdot H_{i,5}, \omega 1 \cdot H_{i,6}, s)_3$$

SETUP FOR POLAR PLOTS

$$f_i := arg(AB_i) \qquad Mg_i := |AB_i| \qquad g1_i := |ABL_i|$$

$$f1_i := arg(ABL_i) \qquad Mg2_i := |ABH_i| \qquad \phi2_i := arg(ABH_i)$$

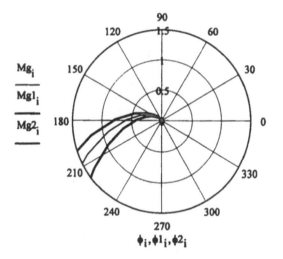

FIGURE 53.   Nyquist polar plot.

In Figure 53, note phase margin: <45°; Cx = 80pF.

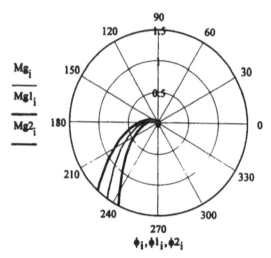

$$\frac{Mg_i}{\overline{\overline{Mg1_i}}}$$
$$\overline{\overline{Mg2_i}}$$

$\phi_i, \phi1_i, \phi2_i$

FIGURE 54. Nyquist polar plot.

In Figure 54, note improved phase margin Cx = 200pF.

TOLERANCE ANALYSIS OF AN A-TO-D CIRCUIT

(Temperature sensing test circuit):

The analog input to an A-to-D converter is next analyzed to determine functional test limits. In this application the test technician was required to read a set of 8 binary LEDs to determine the correct level of output.[*] In a less than state-of-the-art configuration, the binary output of the A-to-D would be read by test software. In either event, the correct range of digital outputs must be calculated.

As Mathcad has no function for direct decimal to binary conversion, the following explanation of one way to do it will facilitate the understanding of the calculation sequences used in the analysis to follow. The example converts 25d to 11001b.

---

[*] 1997 manual test procedure.

DECIMAL TO 8-BIT BINARY CONVERSION: $n := 1 .. 8$

$$re_n := 25 \qquad re_{n+1} := floor\left(\frac{re_n}{2}\right)$$

$re^T = [25\ 12\ 6\ 3\ 1\ 0\ 0\ 0\ 0]$ $ba_n := re_n - 2 \cdot re_{n+1}$

$ba^T = [1\ 0\ 0\ 1\ 1\ 0\ 0\ 0]$ $bb := \sum_n ba_n \cdot 10^{n-1}$ $bb = 11001$

TEST CIRCUIT

R3 thru R6 simulate an RTD at various temperatures, and are manually switched in and out. These resistors were located on the manual test box; R1, R2, the CMOS switch, and the A-to-D were located on the circuit card being tested.

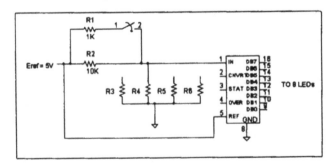

$R2 := 10$ $R1 := 1$ $u := 1 .. 2$ (CMOS switch position)

$w := 1 .. 4$ (RTD switch selector)

$R3 := 75.79$ $R4 := 7.355$ $R5 := 2.25$ $R6 := 0.395$

$Rx := (R3\ R4\ R5\ R6)^T$

$$\text{Rin}_u := \text{if}\left(u = 1, R2, \frac{R1 \cdot R2}{R1 + R2}\right) \qquad \text{Rin}^T = [10 \quad 0.909]$$

$$\text{Eref} := 5 \quad D := 255 \quad (\text{8-bit A-to-D}) \qquad G(\text{Rin}, \text{Rx}) := \frac{\text{Rx}}{\text{Rin} + \text{Rx}}$$

$$Va_{u,w} := \text{Eref} \cdot G(\text{Rin}_u, \text{Rx}_w) \qquad Va = \begin{bmatrix} 4.417 & 2.119 & 0.918 & 0.19 \\ 4.941 & 4.45 & 3.561 & 1.514 \end{bmatrix}$$

Va is the input voltage to the A-to-D on a scale of 5V = 255d.

Convert to 8-bit decimal <255:

$$nv_{u,w} := D \cdot G(\text{Rin}_u, \text{Rx}_w)$$

$$nv = \begin{bmatrix} 225.28 & 108.07 & 46.84 & 9.69 \\ 251.98 & 226.95 & 181.62 & 77.24 \end{bmatrix} \qquad (\text{Eref ratios out})$$

Round to nearest integer:

$$nv_{u,w} := \text{floor}(D \cdot G(\text{Rin}_u, \text{Rx}_w) + 0.5)$$

$$nv = \begin{bmatrix} 225 & 108 & 47 & 10 \\ 252 & 227 & 182 & 77 \end{bmatrix}$$

$$\qquad\qquad\qquad\qquad\qquad\qquad\qquad\qquad \text{Rin} \quad \text{Rx} \quad \text{A/D}$$

$$\text{Tr} := 0.02 \quad \text{Tad} := \frac{1}{D} \quad (\text{1 LSB error}) \qquad T := \begin{bmatrix} -\text{Tr} & -\text{Tr} & -\text{Tad} \\ \text{Tr} & \text{Tr} & \text{Tad} \end{bmatrix}$$

$$k := 1..2 \quad \text{Nc} := 3 \quad p := 1..\text{Nc} \quad \text{Sen} := (-1 \quad 1 \quad 1)^T$$

$$\text{(Sensitivity signs by inspection)}$$

$$M_{k,p} := \text{if}(k = 1, \text{if}(\text{Sen}_p > 0, 1 + T_{1,p}, 1 + T_{2,p}),$$
$$\text{if}(\text{Sen}_p > 0, 1 + T_{2,p}, 1 + T_{1,p}))$$

$$M = \begin{bmatrix} 1.02 & 0.98 & 0.996 \\ 0.98 & 1.02 & 1.004 \end{bmatrix}$$

Convert to EVA decimal count:

$$\text{nvlo}_{u,w} := \text{floor}(D \cdot M_{1,3} \cdot G(\text{Rin}_u \cdot M_{1,1}, Rx_w \cdot M_{1,2}) + 0.5)$$

$$\text{nvlo} = \begin{bmatrix} 223 & 105 & 45 & 9 \\ 251 & 225 & 179 & 75 \end{bmatrix}$$

$$\text{nvhi}_{u,w} := \text{floor}(D \cdot M_{2,3} \cdot G(\text{Rin}_u \cdot M_{2,1}, Rx_w \cdot M_{2,2}) + 0.5)$$

$$\text{nvhi} = \begin{bmatrix} 227 & 111 & 49 & 10 \\ 253 & 229 & 184 & 80 \end{bmatrix}$$

Display arrays: neva := stack(nvlo, nvhi) Rin := stack(Rin, Rin)

$$\text{neva} = \begin{bmatrix} 223 & 105 & 45 & 9 \\ 251 & 225 & 179 & 75 \\ 227 & 111 & 49 & 10 \\ 253 & 229 & 184 & 80 \end{bmatrix} \qquad \text{Rin} = \begin{bmatrix} 10 \\ 0.909 \\ 10 \\ 0.909 \end{bmatrix}$$

Convert to 8-bit binary for LED readout or software per procedure on page 131.

$$\text{re}_{a,w} := \text{neva}_{1,w} \quad \text{re}_{a+1,w} := \text{floor}\left(\frac{\text{re}_{a,w}}{2}\right)$$

$$\text{bl}_w := \sum_a \left(\text{re}_{a,w} - 2 \cdot \text{re}_{a+1,w}\right) \cdot 10^{a-1} \quad \text{row 1}$$

$$re_{n,w} := neva_{2,w} \quad re_{n+1,w} := floor\left(\frac{re_{n,w}}{2}\right)$$

$$b2_w := \sum_n \left(re_{n,w} - 2 \cdot re_{n+1,w}\right) \cdot 10^{n-1} \quad \text{row 2}$$

$$re_{n,w} := neva_{3,w} \quad re_{n+1,w} := floor\left(\frac{re_{n,w}}{2}\right)$$

$$b3_w := \sum_n \left(re_{n,w} - 2 \cdot re_{n+1,w}\right) \cdot 10^{n-1} \quad \text{row 3}$$

$$re_{n,w} := neva_{4,w} \quad re_{n+1,w} := floor\left(\frac{re_{n,w}}{2}\right)$$

$$b4_w := \sum_n \left(re_{n,w} - 2 \cdot re_{n+1,w}\right) \cdot 10^{n-1} \quad \text{row 4}$$

$$bin := augment(b1, augment(b2, augment(b3, b4)))$$

Form test procedure array, Rx in first row, Rin on right:
$$bin := augment(Rx, bin)$$

$$bin^T = \begin{bmatrix} 75.79 & 7.355 & 2.25 & 0.395 \\ 11011111 & 1101001 & 101101 & 1001 \\ 11111011 & 11100001 & 10110011 & 1001011 \\ 11100011 & 1101111 & 110001 & 1010 \\ 11111101 & 11100101 & 10111000 & 1010000 \end{bmatrix}$$

$$\text{Rin} = \begin{bmatrix} 10 \\ 0.909 \\ 10 \\ 0.909 \end{bmatrix} \quad \begin{matrix} \text{EVL} \\ \\ \text{EVH} \end{matrix}$$

Example:

$$\text{Rx} = 0.395 \text{ and Rin} = 0.909$$

$$\text{EVL} = 1001011b = 75d = 4Bh$$

$$\text{EVH} = 1010000b = 80d = 50h$$

To aid the test technician, provide a range of binary numbers between high and low in the test procedure.

Example:

$$n := 1 .. 8 \qquad \text{Nr} := 75 .. 80 \qquad \text{ra}_{n,\text{Nr}} := \text{Nr}$$

$$\text{ra}_{n+1,\text{Nr}} := \text{floor}\left(\frac{\text{ra}_{n,\text{Nr}}}{2}\right)$$

$$\text{ba}_{\text{Nr}} := \sum_{n}\left(\text{ra}_{n,\text{Nr}} - 2 \cdot \text{ra}_{n+1,\text{Nr}}\right) \cdot 10^{n-1}$$

ba_{Nr}
1001011
1001100
1001101
1001110
1001111
1010000

NS MF10 SWITCHED CAPACITOR FILTER — MODE 3, BPF (LTC1060)

This device is manufactured by two major semiconductor manufacturers. This sequence will show a design procedure as well as the tolerance analysis.

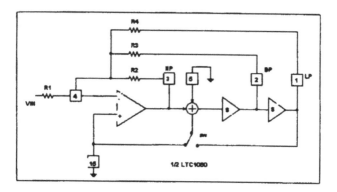

Design center frequency: 640 Hz, with an overall Q for both stages of 100; gain H = 1.

$$K := 10^3 \qquad fo := 640 \qquad BW := 6.4 \qquad Qa := \frac{fo}{BW} \qquad Qa = 100$$

Two stages:

$$Qb := Qa\sqrt{\sqrt{2}-1} \qquad\qquad Qb = 64.359$$

Use Qb to design for overall Q of 100.

Design: fc := 51000   (fc = switching frequency)   H := 1

$$wc := \frac{2 \cdot \pi \cdot fc}{50} \quad wo := 2 \cdot \pi \cdot fo$$

Choose R1:   R1 := 412 · K

Calculate R2 thru R4:

$$R2 := \frac{H \cdot R1 \cdot wo}{Qb \cdot wc} \quad R3 := \frac{wc \cdot Qb \cdot R2}{wo} \quad R4 := R2 \cdot \left(\frac{wc}{wo}\right)^2$$

$$R2 = 4.017 \circ K \quad R3 = 412 \circ K \quad R4 = 10.202 \circ K$$

R5 := R1   R6 := R2   R7 := R3   R8 := R4

(2nd stage resistors identical)

$$BF := 600 \quad LF := 680 \quad DF := 1 \quad i := 1.. \frac{LF - BF}{DF} + 1$$

$$F_i := BF + DF \cdot (i - 1) \quad s_i := 2 \cdot \pi \cdot F_i \cdot \sqrt{-1}$$

Resistor tolerance:

$$Tr := 0.01$$

Switching frequency tolerance:

$$Tf := 0.037$$

$$G(R1, R2, R3, R4, R5, R6, R7, R8, wc, s) :=$$

$$
\begin{vmatrix}
Tfa \leftarrow \dfrac{\dfrac{wc \cdot R2 \cdot s_i}{R1}}{\left(s_i\right)^2 + \dfrac{wc \cdot R2 \cdot s_i}{R3} + wc^2 \dfrac{R2}{R4}} \\[3em]
Tfb \leftarrow \dfrac{\dfrac{wc \cdot R6 \cdot s_i}{R5}}{\left(s_i\right)^2 + \dfrac{wc \cdot R6 \cdot s_i}{R7} + wc^2 \dfrac{R6}{R8}} \\[3em]
Tf \leftarrow Tfa \cdot Tfb \\[0.5em]
Tf
\end{vmatrix}
$$

$$Voi := G(R1, R2, R3, R4, R5, R6, R7, R8, wc, s)$$

$$Nc := 9 \qquad N := 2000 \qquad k := 1 .. N \qquad w := 1 .. Nc$$

$$
T := \begin{bmatrix}
-Tr & -Tr & -Tr & -Tr & -Tr & -Tr & -Tr & -Tr & -Tf \\
Tr & Tr & Tr & Tr & Tr & Tr & Tr & Tr & Tf
\end{bmatrix}
$$

$$Tn_{w,k} := (T_{2,w} - T_{1,w}) \cdot rnd(1) + T_{1,w} + 1$$

Vm is obtained in the usual manner:

$$Vm_{k,1} = G(R1 \cdot Tn_{k,1}, R2 \cdot Tn_{k,2}, \ldots, wc \cdot Tn_{k,9}, s)$$

$$Vmax_i := max(Vm^{<i>})$$

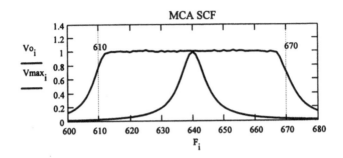

FIGURE 55. MCA SCF. (N = 2000).

The markers at 610 and 670 Hz were obtained from an FMCA analysis.

400-HZ FULL-WAVE RECTIFIER CIRCUIT

Transient analysis to dc equivalent circuit for EVA/RSS analysis is facilitated in this circuit because the HWR diodes do not have to be modeled. The output at Vo2 is not affected by the diode drop of D1.

This circuit is used in an aircraft ac power control system. The 0.25 Vrms input below is from a step-down current transformer when the ac bus current is 250 Amps. The ac input is then converted to +5 Vdc for monitoring purposes. Vo3 is applied to under and over-current window comparators.

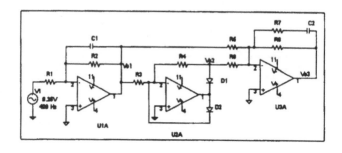

$K := 10^3$ $nF := 10^{-9}$ $uF := 10^{-6}$ $mV := 10^{-3}$ $Vpk := 1$

$R1 := 630$ $R2 := 14 \cdot K$ $C1 := 1 \cdot nF$ $R3 := 20 \cdot K$

$R4 := 20 \cdot K$ $R5 := 20 \cdot K$

$R6 := 10 \cdot K$ $R7 := 100$ $R8 := 20 \cdot K$ $C2 := 1.1 \cdot uF$

$$Vol := \frac{0.25 \cdot R2 \cdot \sqrt{2}}{R1}$$ $Vol = 7.857 \circ Vpk$ $F := 400$ $T := \frac{1}{F}$

$Tn := 3$ (Plot Tn periods) $t := 0, \frac{T}{50} .. Tn \cdot T$ $\omega := 2 \cdot \pi \cdot F$

Show half-wave to full-wave conversion:

Input via R5 (gain of 1): $f5(t) := Vol \cdot \sin(\omega \cdot t + \pi)$

$$f6(t) := \begin{vmatrix} 0 \text{ if } 0 \le t \le \dfrac{T}{2} & \text{f6(t) is the input via R6 (gain of 2)} \\ 2 \cdot Vol \cdot \sin(\omega \cdot t) \text{ if } t > \dfrac{T}{2} \\ f6(t - T) \text{ if } t > T \end{vmatrix}$$

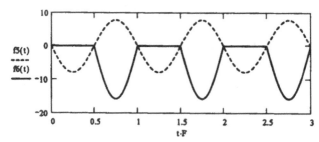

FIGURE 56. Halfwave rectifier.

This solid plot represents the current in R6, while the dashed line represents the current in R5. When these two currents are summed by U3A, the output is a full-wave rectifier before filtering.

Invert and filter full-wave (f5 + f6); get Fourier cofficients; use 10 harmonics: $k := 1 .. 10$

$$f(t) := Vo1 \cdot |\sin(\omega \cdot t)| \qquad A_k := \frac{2}{T} \cdot \int_0^T f(t) \cdot \cos(k \cdot \omega \cdot t)\, dt$$

$$B_k := \frac{2}{T} \cdot \int_0^T f(t) \cdot \sin(k \cdot \omega \cdot t)\, dt \qquad C_k := \sqrt{\left(A_k\right)^2 + \left(B_k\right)^2}$$

$$\phi_k := -\operatorname{atan}\left(\frac{B_k}{A_k}\right) \qquad A_0 := \frac{1}{T} \cdot \int_0^T f(t)\, dt$$

$$H2(s) = \frac{s \cdot R7 \cdot R8 \cdot C2 + R8}{s \cdot C2 \cdot (R7 + R8) + 1} \cdot \frac{1}{R5} \qquad \text{(U3A transfer function)}$$

$$N1 := \frac{R7 \cdot R8 \cdot C2}{R5} \qquad N0 := \frac{R8}{R5}$$

$$D1 := C2 \cdot (R7 + R8) \quad D0 := 1 \quad G_0 := \frac{N0}{D0} \quad G_0 = 1$$

$$A_0 \cdot G_0 = 5.002 \qquad \frac{2 \cdot Vol}{\pi} = 5.002$$

Calculate the transfer function harmonic amplitudes and phase angles:

$$G_k := \sqrt{\frac{N0^2 + (k \cdot \omega \cdot N1)^2}{D0^2 + (k \cdot \omega \cdot D1)^2}}$$

$$\theta_k := \operatorname{atan}\left(\frac{k \cdot \omega \cdot N1}{N0}\right) - \operatorname{atan}\left(\frac{k \cdot \omega \cdot D1}{D0}\right) + \pi$$

$$h(t) := A_0 \cdot G_0 + \sum_k 2 \cdot |C_k| \cdot G_k \cdot \cos\left(k \cdot \omega \cdot t + \phi_k + \theta_k\right)$$

(Inverse Fourier Transform)

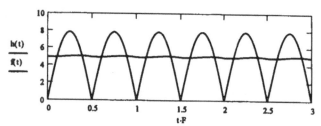

FIGURE 57.   Fullwave rectifier and dc avg.

$g(t) := h(t) - A_0 \cdot G_0$ (Put "scope" in AC mode to show pk-to-pk ripple)

$Va := 0.07 \quad Vb := -0.066 \quad Vrpk := Va - Vb \quad Vrpk = 136 \circ mV$

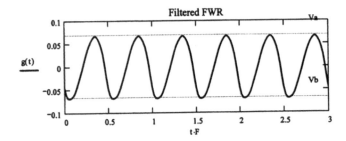

FIGURE 58. Filtered FWR (ripple).

Now we are in a position to perform a RSS/EVA on the 400-Hz
FWR dc equation. 400-Hz FWR circuit – dc output of U3A (Assumes
R7 = inf.). (See page 140 for resistor values and the Appendix for
derivation of the dc equation.)

$$\text{Vin} := 0.25 \cdot \text{Vrms} \qquad \text{Nc} := 7 \qquad p := 1 .. \text{Nc} \qquad \text{dpf} := 0.0001$$

$$Q := \text{dpf} \cdot \text{identity (Nc)} + 1$$

Dc equation:

$$G(\text{R1}, \text{R2}, \text{R3}, \text{R4}, \text{R6}, \text{R8}, \text{Vin}) := \frac{\text{Vin} \cdot \text{R2} \cdot \text{R4} \cdot \text{R8} \cdot \sqrt{2}}{\pi \cdot \text{R1} \cdot \text{R3} \cdot \text{R6}}$$

$$\text{Vo} := G(\text{R1}, \text{R2}, \text{R3}, \text{R4}, \text{R6}, \text{R8}, \text{Vin}) \qquad \text{Vo} = 5.002$$

$$\text{Vr}_p := G(\text{R1} \cdot Q_{p,1}, \text{R2} \cdot Q_{p,2}, \text{R3} \cdot Q_{p,3}, \text{R4} \cdot Q_{p,4},$$
$$\text{R6} \cdot Q_{p,5}, \text{R8} \cdot Q_{p,6}, \text{Vin} \cdot Q_{p,7})$$

$$\text{Sen}_p := \left(\frac{\text{Vr}_p}{\text{Vo}} - 1 \right) \cdot \frac{1}{\text{dpf}} \qquad \text{Sen}^T = [-1 \ \ 1 \ \ -1 \ \ 1 \ \ -1 \ \ 1 \ \ 1]$$

$$Tr := 0.001 \quad Tv := 0 \quad T := \begin{bmatrix} -Tr & -Tr & -Tr & -Tr & -Tr & -Tr & -Tv \\ Tr & Tr & Tr & Tr & Tr & Tr & Tv \end{bmatrix}$$

$$k := 1..2$$

$$M_{k,p} := if(k = 1, if(Sen_p > 0, 1 + T_{1,p}, 1 + T_{2,p}),$$
$$\qquad\qquad if(Sen_p > 0, 1 + T_{2,p}, 1 + T_{1,p}))$$

$$Vev_k := G(R1 \cdot M_{k,1}, R2 \cdot M_{k,2}, R3 \cdot M_{k,3}, R4 \cdot M_{k,4},$$
$$\qquad\qquad R6 \cdot M_{k,5}, R8 \cdot M_{k,6}, Vin \cdot M_{k,7})$$

$$Vrss_k := Vo \cdot \left[1 + (-1)^k \cdot \sqrt{\sum_p \left[Sen_p \cdot \left(M_{k,p} - 1 \right) \right]^2} \right]$$

$$Vev = \begin{bmatrix} 4.972 \\ 5.032 \end{bmatrix} \qquad Vrss = \begin{bmatrix} 4.99 \\ 5.014 \end{bmatrix}$$

In terms of ac bus current, the EVA is:

$$sf := \frac{250}{5} \qquad sf \cdot Vev = \begin{bmatrix} 248.59 \\ 251.59 \end{bmatrix}$$

CONFIDENCE INTERVALS FOR 3σ RSS*

Form N vector for variable number of samples.

$$N := (100 \ 200 \ 500 \ 1000 \ 2000 \ 5000 \ 10000 \ 20000 \ 50000)^T$$

$$k := 1..rows(N) \qquad u := 1..2$$

---

* See page 81, Reference 7.

Use two confidence levels:

$$\alpha = 90\% \text{ and } 95\% \qquad \alpha := (0.9 \quad 0.95)^T \qquad p := \frac{1+\alpha}{2}$$

(qchisq is Mathcad's χ^2 statistical function. Calculate A and B factors:

$$A_{k,s} := \sqrt{\frac{N_k}{\text{qchisq}(p_s, N_k)}} \qquad B_{k,s} := \sqrt{\frac{N_k}{\text{qchisq}(1-p_s, N_k)}}$$

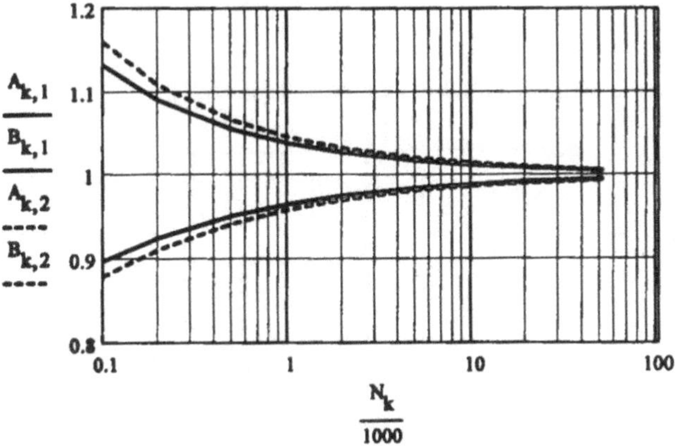

FIGURE 59. 3s 90% and 95% confidence intervals.

Example: N = 1000. 95% confidence level:

$$N_4 = 1000 \qquad A_{4,2} = 0.958 \qquad B_{4,2} = 1.046$$

Assume the MCA of a circuit gave an average output of Vavg := 20 and Vrss := 1.5

Then with 95% confidence the true 3σ value is between:

$$\text{Vavg} + \text{Vrss} \cdot A_{4,2} = 21.437 \quad \text{and} \quad \text{Vavg} + \text{Vrss} \cdot B_{4,2} = 21.569$$

$$\text{Avc} := \text{Vrss} \cdot (B_{4,2} - A_{4,2}) \qquad \Delta\text{Vc} = 0.132$$

Note how the confidence interval ΔVc decreases with increasing N. Hence the reason for using a high value of N.

CONFIDENCE INTERVAL FOR THE MEAN

$$C_{k,a} := \frac{qt(p_a, N_k)}{\sqrt{N_k}}$$

(qt is Mathcad's Student's t statistical function.)

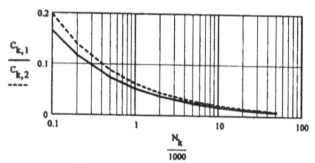

FIGURE 60. Mean 90% and 95% C.I.s.

Example: $N = 500$ \qquad $\text{Vavg} = 20$ \qquad $\text{Vrss} = 1.5$

$$C_{3,2} = 0.088 \qquad C_{3,1} = 0.074$$

With 95% confidence the true mean is between

$$\text{Vavg} - \text{Vrss} \cdot C3,2 = 19.868 \quad \text{and} \quad \text{Vavg} + \text{Vrss} \cdot C3,2 = 20.132$$

(95% is the confidence level, ΔVc is the confidence interval.)

With 90% confidence, the true mean is between

$$Vavg - Vrss \cdot C_{3,1} = 19.889 \quad \text{and} \quad Vavg + Vrss \cdot C_{3,1} = 20.111$$

LARGE CIRCUIT

This is an example of a large (14th order) circuit analyzed using Mathcad. It should be noted that with a matrix this large, the memory of some PC's may be taxed to the limit just for the nominal output. MCA can be accomplished with a small value of N.

7th Order lowpass elliptical filter (buffers used at V7 and V13).
$Ao := 10^6;\ nF := 10^{-9}$.

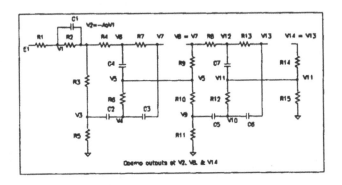

Observe outputs at V2, V8, & V14

$$K := 10^3 \quad R1 := 19.6 \cdot K \quad R2 := .196 \cdot K \quad R3 := 1 \cdot K$$

$$C1 := 2.67 \cdot nF \quad \text{(Sets break point)}$$

$$C2 := C1 \quad C3 := C1 \quad C4 := C1 \quad C5 := C1 \quad C6 := C1 \quad C7 := C1$$

$$R4 := 147 \cdot K \quad R5 := 71.5 \quad R6 := 37.4 \cdot K \quad R7 := 154 \cdot K$$

$$R8 := 110 \cdot K \quad R9 := 260$$

$$R10 := 740 \qquad R11 := 402 \qquad R12 := 27.4 \cdot K$$

$$R13 := 110 \cdot K \qquad R14 := 40 \qquad R15 := 960$$

$$Bf := 2 \qquad ND := 2 \qquad PD := 50$$

$$i := 1 .. ND \cdot PD + 1 \qquad\qquad L_i := BF + \frac{(i-1)}{PD}$$

$$F_i := 10^{L_i} \qquad s_i := 2 \cdot \pi \cdot F_i \cdot \sqrt{-1} \qquad db(x) := 20 \cdot \log(|x|)$$

Construct matrix elements from schematic using the mnemonic method:

$$A11_i := \frac{1}{R1} + \frac{s_i \cdot R2 \cdot C1 + 1}{R2} \qquad A12_i := -\left(\frac{s_i \cdot R2 \cdot C1 + 1}{R2} \right)$$

$$A32 := \frac{-1}{R3} \qquad A33_i := \frac{1}{R3} + \frac{1}{R5} + s_i \cdot C2 \qquad A34_i := -s_i \cdot C2$$

$$A43_i := A34_i \qquad A44_i := s_i \cdot C2 + s_i \cdot C3 + \frac{1}{R6} \qquad A45 := \frac{-1}{R6}$$

$$A47_i := -s_i \cdot C3 \qquad A54 := A45 \qquad A55_i := s_i \cdot C4 + \frac{1}{R6} + \frac{1}{R9} + \frac{1}{R10}$$

$$A56_i := -s_i \cdot C4 \qquad A58 := \frac{-1}{R9} \qquad A59 := \frac{-1}{R10}$$

$$A62 := \frac{-1}{R4} \qquad A65_i := A56_i \qquad A66_i := s_i \cdot C4 + \frac{1}{R4} + \frac{1}{R7}$$

$$A67 := \frac{-1}{R7} \qquad A74_i := A47_i \qquad A76 := A67$$

$$A77_i := s_i \cdot C3 + \frac{1}{R7} \qquad A88 := 1 \qquad A87 := -1 \qquad (V8 = V7)$$

$$A95 := \frac{-1}{R10} \qquad A99_i := s_i \cdot C5 + \frac{1}{R11} + \frac{1}{R10} \qquad A9\_10_i := -s_i \cdot C5$$

$$A10\_9_i := A9\_10_i \qquad A10\_10_i := s_i \cdot C5 + s_i \cdot C6 + \frac{1}{R12}$$

$$A10\_11 := \frac{-1}{R12} \qquad A10\_13_i := -s_i \cdot C6 \qquad A11\_10 := A10\_11$$

$$A11\_11_i := s_i \cdot C7 + \frac{1}{R12} + \frac{1}{R14} + \frac{1}{R15} \qquad A11\_12_i := -s_i \cdot C7$$

$$A11\_14 := \frac{-1}{R14} \qquad A12\_12_i := s_i \cdot C7 + \frac{1}{R8} + \frac{1}{R13}$$

$$A12\_13 := \frac{-1}{R13} \qquad A12\_11_i := A11\_12_i \qquad A12\_8 := \frac{-1}{R8}$$

$$A13\_10_i := A10\_13_i \qquad A13\_12 := A12\_13$$

$$A13\_13_i := s_i \cdot C6 + \frac{1}{R13} \qquad A14\_14 := 1 \qquad A14\_13 := -1$$

$$A14\_11 := A11\_14$$

Form sub-matrices:

$$
Aa_i := \begin{bmatrix}
A11_i & A12_i & 0 & 0 & 0 & 0 & 0 \\
Ao & 1 & 0 & 0 & 0 & 0 & 0 \\
0 & A32 & A33_i & A34_i & 0 & 0 & 0 \\
0 & 0 & A43_i & A44_i & A45 & 0 & A47_i \\
0 & 0 & 0 & A54 & A55_i & A56_i & 0 \\
0 & A62 & 0 & 0 & A65_i & A66_i & A67 \\
0 & 0 & 0 & A74_i & 0 & A76 & A77_i
\end{bmatrix}
$$

$$
Ab_i := \begin{bmatrix}
0 & 0 & 0 & 0 & 0 & 0 & 0 \\
0 & 0 & 0 & 0 & 0 & 0 & 0 \\
0 & 0 & 0 & 0 & 0 & 0 & 0 \\
0 & 0 & 0 & 0 & 0 & 0 & 0 \\
A58 & A59 & 0 & 0 & 0 & 0 & 0 \\
0 & 0 & 0 & 0 & 0 & 0 & 0 \\
0 & 0 & 0 & 0 & 0 & 0 & 0
\end{bmatrix}
\qquad
Ac_i := \begin{bmatrix}
0 & 0 & 0 & 0 & 0 & 0 & -1 \\
0 & 0 & 0 & 0 & A95 & 0 & 0 \\
0 & 0 & 0 & 0 & 0 & 0 & 0 \\
0 & 0 & 0 & 0 & 0 & 0 & 0 \\
0 & 0 & 0 & 0 & 0 & 0 & 0 \\
0 & 0 & 0 & 0 & 0 & 0 & 0 \\
0 & 0 & 0 & 0 & 0 & 0 & 0
\end{bmatrix}
$$

$$
Ad_i := \begin{bmatrix}
1 & 0 & 0 & 0 & 0 & 0 & 0 \\
0 & A99_i & A9\_10_i & 0 & 0 & 0 & 0 \\
0 & A10\_9_i & A10\_10_i & A10\_11 & 0 & A10\_13_i & 0 \\
0 & 0 & A11\_10 & A11\_11_i & A11\_12_i & 0 & A11\_14 \\
A12\_8 & 0 & 0 & A12\_11 & A12\_12_i & A12\_13 & 0 \\
0 & 0 & A13\_10_i & 0 & A13\_12 & A13\_13_i & 0 \\
0 & 0 & 0 & 0 & 0 & -1 & 1
\end{bmatrix}
$$

Assemble the sub-matrices:

$$AT_i := augment(Aa_i, Ab_i) \qquad AB_i := augment(Ac_i, Ad_i)$$

$$A_i := stack(AT_i, AB_i)$$

$$B_i := \begin{bmatrix} \dfrac{1}{R1} & 0 & 0 & 0 & 0 & 0 & 0 & 0 & 0 & 0 & 0 & 0 & 0 & 0 \end{bmatrix}^T$$

$$C_i := \text{lsolve}\left(A_i, B_i\right) \qquad Vo_i := db\left[\left(C_i\right)_{14}\right]$$

$$Va_i := -210 \cdot \log\left(\frac{F_i}{1000}\right)$$ (Asymptote of rolloff; = −210db/ decade with 7th order circuit.)

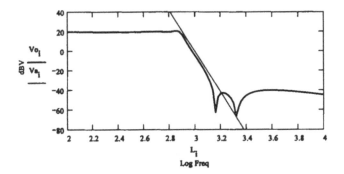

FIGURE 61.   Lowpass elliptical filter.

An analysis package called MATLAB®* (for MATrix LABoratory) is optimized for matrix analysis and can easily handle this circuit. See the companion book, *Tolerance Analysis of Electronic Circuits Using MATLAB*, published by CRC Press LLC.

---

* MATLAB ia a registered trademark of The MathWorks, Inc.

COMPARATOR CIRCUIT

Tolerance analysis of input trip levels. V2 low (Q1 on) when pin2 > pin3;
Rc = collector resistance.

$$Rc := [10 \quad 10^8]^T \qquad K := 10^3 \qquad mV := 10^{-3}$$

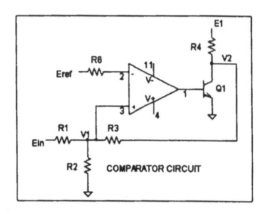

$$Eref := 5 \qquad E1 := 5 \qquad R1 := 13.16 \cdot K \qquad R2 := 20 \cdot K$$

$$R3 := 250 \cdot K \qquad R4 := 3.3 \cdot K$$

$$G(R1, R2, R3, R4, Rc, Ein, E1) :=$$

$$\begin{vmatrix} A \leftarrow \begin{bmatrix} \dfrac{1}{R1} + \dfrac{1}{R2} + \dfrac{1}{R3} & \dfrac{-1}{R3} \\[2ex] \dfrac{-1}{R3} & \dfrac{1}{Rc} + \dfrac{1}{R4} + \dfrac{1}{R3} \end{bmatrix} \\[4ex] B \leftarrow \begin{bmatrix} \dfrac{Ein}{R1} & \dfrac{E1}{R4} \end{bmatrix}^T \\[2ex] C \leftarrow \text{lsolve}(A, B) \\[1ex] C_1 \end{vmatrix}$$

Create a ramp input:

$$e := 1 .. 201 \qquad Ein_e := \frac{e - 1}{20}$$

$$Vh1_e := G(R1, R2, R3, R4, Rc_1, Ein_e, E1) \qquad M1 := 8.552$$

$$Vh2_e := G(R1, R2, R3, R4, Rc_2, Ein_e, E1)$$

$$Vc_e := \text{if}(Vh1_e > Eref, Vh2_e, Vh1_e) \qquad M2 := 8.288$$

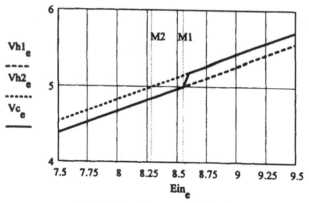

FIGURE 62. Comparator trip levels.

Output (V2) goes high at an input of M1 and goes low at M2. Total hysteresis is M1 − M2 = 0.264.

Need to find input levels where V1 = 5V. Ein is now dependent and V1 is independent. Swap dependent and independent variables in first equation (first row of matrix).   V1 := 5

$$H(R1, R2, R3, R4, Rc, V1, E1) :=$$

$$\begin{vmatrix} A \leftarrow \begin{bmatrix} \dfrac{-1}{R1} & \dfrac{-1}{R3} \\[2ex] \dfrac{-1}{R3} & \dfrac{1}{Rc} + \dfrac{1}{R4} + \dfrac{1}{R3} \end{bmatrix} \\[4ex] B \leftarrow \begin{bmatrix} -V1 \cdot \left(\dfrac{1}{R1} + \dfrac{1}{R2} + \dfrac{1}{R3} \right) & \dfrac{E1}{R4} \end{bmatrix}^T \\[3ex] C \leftarrow lsolve(A, B) \\[1ex] C_1 \end{vmatrix}$$

$$u := 1..2 \qquad Vin_u := H(R1, R2, R3, R4, Rc_u, V1, E1)$$

$$Vin = \begin{bmatrix} 8.552 \\ 8.288 \end{bmatrix} \qquad hyst := Vin_1 - Vin_2$$

$$hyst = 0.265 \qquad Nc := 7 \qquad p := 1..Nc \qquad dpf := 0.0001$$

$$Q := dpf \cdot identity(Nc) + 1 \qquad k := 1..2$$

$$Vinr_{p,k} := H(R1 \cdot Q_{p,1},\ R2 \cdot Q_{p,2}, R3 \cdot Q_{p,3}, R4 \cdot Q_{p,4},$$
$$Rc_k \cdot Q_{p,5}, V1 \cdot Q_{p,6}, E1 \cdot Q_{p,7})$$

$$Sen_{p,k} := \left(\frac{Vinr_{p,k}}{Vin_k} - 1 \right) \cdot \frac{1}{dpf} \qquad Tr := 0.001 \qquad Trc := 0.1$$

$$Tv := 0.002 \qquad Te := 0.1$$

R1 thru R3 are precision metal film; pullup R4 is 10% carbon. Eref tolerance = Tv, Vcc tolerance = Te. Two sets of sensitivities this time, i.e., circuit undergoes a change when Q1 goes on and off. Hence two EVA/RSS analyses. Note sign change of sensitivities for Q1 on and off.

$$T := \begin{bmatrix} -Tr & -Tr & -Tr & -Trc & 0 & -Tv & -Te \\ Tr & Tr & Tr & Trc & 0 & Tv & Te \end{bmatrix}$$

$$Sen = \begin{bmatrix} 0.415 & 0.396 \\ -0.385 & -0.397 \\ -0.031 & 5.391 \cdot 10^{-4} \\ 9.268 \cdot 10^{-5} & -2.673 \cdot 10^{-4} \\ -9.478 \cdot 10^{-5} & -1.042 \cdot 10^{-6} \\ 1 & 1.031 \\ -9.297 \cdot 10^{-5} & -0.031 \end{bmatrix}$$

$$M1_{k,p} := \text{if}(k = 1, \text{if}(Sen_{p,1} < 0, 1 + T_{2,p}, 1 + T_{1,p}),$$
$$\text{if}(Sen_{p,1} < 0, 1 + T_{1,p}, 1 + T_{2,p}))$$

$$M2_{k,p} := \text{if}(k = 1, \text{if}(Sen_{p,2} < 0, 1 + T_{2,p}, 1 + T_{1,p}),$$
$$\text{if}(Sen_{p,2} < 0, 1 + T_{1,p}, 1 + T_{2,p}))$$

$$Vev_{k,1} := H(R1 \cdot M1_{k,1}, R2 \cdot M1_{k,2}, R3 \cdot M1_{k,3}, R4 \cdot M1_{k,4},$$
$$Rc_1 \cdot M1_{k,5}, V1 \cdot M1_{k,6}, E1 \cdot M1_{k,7})$$

$$Vev_{k,2} := H(R1 \cdot M2_{k,1}, R2 \cdot M2_{k,2}, R3 \cdot M2_{k,3}, R4 \cdot M2_{k,4},$$
$$Rc_2 \cdot M2_{k,5}, V1 \cdot M2_{k,6}, E1 \cdot M2_{k,7})$$

$$Vrss_{k,1} := Vin_1 \cdot \left[1 + (-1)^k \cdot \sqrt{\sum_p \left[Sen_{p,1} \cdot \left(M1_{k,p} - 1 \right) \right]^2} \right]$$

$$Vrss_{k,2} := Vin_2 \cdot \left[1 + (-1)^k \cdot \sqrt{\sum_p \left[Sen_{p,2} \cdot \left(M2_{k,p} - 1 \right) \right]^2} \right]$$

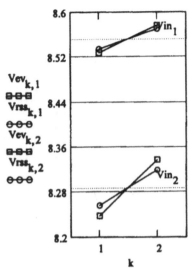

FIGURE 63. EVA/RSS of trip levels.

$$Vev = \begin{bmatrix} 8.528 & 8.238 \\ 8.577 & 8.338 \end{bmatrix} \qquad Vrss = \begin{bmatrix} 8.535 & 8.256 \\ 8.57 & 8.319 \end{bmatrix}$$

$$Vin = \begin{bmatrix} 8.552 \\ 8.288 \end{bmatrix} \qquad \text{(Nominal)}$$

Compute deltas:

$$\Delta Vev_{\text{u}} := Vev_{2,\text{u}} - Vev_{1,\text{u}} \qquad \Delta Vrss_{\text{u}} := Vrss_{2,\text{u}} - Vrss_{1,\text{u}}$$

$$\Delta Vev = \begin{bmatrix} 48.74 \\ 99.70 \end{bmatrix} \circ mV \qquad \Delta Vrss = \begin{bmatrix} 35.56 \\ 62.86 \end{bmatrix} \circ mV$$

100-HZ CLOCK GENERATOR

This circuit involves a lengthy series of calculations:

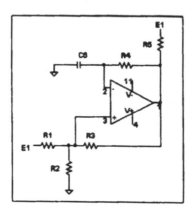

$K := 10^3$ $uF := 10^{-6}$ $ms := 10^{-3}$ $R1 := 100 \cdot K$

$R2 := 100 \cdot K$ $E1 := 5$ $R4 := 5750$ $R5 := 3.3 \cdot K$ $C6 := 1 \cdot uF$

$R3 := 100 \cdot K$

     Vhi is Vc final steady-state *charge* voltage of C6

     Vlo is Vc final steady-state *discharge* = Vce(sat)

     Va is *upper* trigger level

     Vb is *lower* trigger level

$Vlo := 0.25$ (LM339 Vcesat)

$$A := \begin{bmatrix} \dfrac{1}{R1} + \dfrac{1}{R2} + \dfrac{1}{R3} & \dfrac{-1}{R3} \\ \dfrac{-1}{R3} & \dfrac{1}{R3} + \dfrac{1}{R5} \end{bmatrix} \quad B := \begin{bmatrix} \dfrac{E1}{R1} \\ \dfrac{E1}{R5} \end{bmatrix} \quad C := lsolve(A, B)$$

$$C^T = [3.315 \quad 4.946] \qquad Va := C_1 \qquad Vhi := C_2$$

$$Vb := \frac{E1}{1 + R1 \cdot \left(\dfrac{1}{R2} + \dfrac{1}{R3} \right)} + \frac{Vlo}{1 + R3 \cdot \left(\dfrac{1}{R1} + \dfrac{1}{R2} \right)}$$

$$Vb = 1.75 \qquad Rp(a, b) := \frac{a \cdot b}{a + b}$$

Charging time constant τa:

$$Rth = R4 + R5//(R3 + R1//R2) \qquad (// \Rightarrow \text{in parallel})$$

$$Rth := (R4 + Rp(R5, Rp(R3, Rp(R1, R2)))) \quad Rth = 8752.73$$

$$\tau a := Rth \cdot C6$$

Discharge time constant τb:

$$\tau b := R4 \cdot C6$$

Time of 1st half cycle:

$$T1 := \tau a \cdot \ln\left(\frac{Vhi - Vb}{Vhi - Va} \right)$$

Time of 2nd half cycle:

$$T2 := \tau b \cdot \ln\left(\frac{Vlo - Va}{Vlo - Vb} \right)$$

$$T1 = 5.89 \circ ms \quad T2 = 4.11 \circ ms \quad Freq := \frac{1}{T1 + T2} \quad Freq = 100.01$$

Combining these equations into one function would be tedious and error-prone. Use programming to form G-function.

$$G(R1, R2, R3, R4, Rc, Ein, El) :=$$

$$
\begin{vmatrix}
A \leftarrow \begin{bmatrix} \dfrac{1}{R1} + \dfrac{1}{R2} + \dfrac{1}{R3} & \dfrac{-1}{R3} \\[2ex] \dfrac{-1}{R3} & \dfrac{1}{R3} + \dfrac{1}{R5} \end{bmatrix} \\[4ex]
B \leftarrow \begin{bmatrix} \dfrac{El}{R1} & \dfrac{El}{R5} \end{bmatrix}^{T} \\[2ex]
C \leftarrow \text{lsolve}(A, B) \\[1ex]
Vhi \leftarrow C_2 \\[1ex]
Va \leftarrow C_1 \\[1ex]
Vb \leftarrow \dfrac{El}{1 + R1 \cdot \left(\dfrac{1}{R2} + \dfrac{1}{R3} \right)} + \dfrac{Vlo}{1 + R3 \cdot \left(\dfrac{1}{R1} + \dfrac{1}{R2} \right)} \\[4ex]
\tau a \leftarrow C6 \cdot \Big(\big(R4 + Rp(R5, Rp(R3, Rp(R1, R2))) \big) \Big) \\[2ex]
tb \leftarrow R4 \cdot C6 \\[1ex]
Tf \leftarrow \tau a \cdot \ln\left(\dfrac{Vhi - Vb}{Vhi - Va} \right) + \tau b \cdot \ln\left(\dfrac{Vlo - Va}{Vlo - Vb} \right) \\[3ex]
F \leftarrow \dfrac{1}{Tf} \\[2ex]
F
\end{vmatrix}
$$

$$\text{Freq} := G(R1, R2, R3, R4, R5, C6, El) \qquad \text{Freq} = 100.01$$
$$\text{(Nominal design frequency)}$$

For all EVA/RSS/MCA/FMCA analyses, once the G-function is created, the remaining calculations are consistent in form and thus straightforward.

COMPLEMENTARY FEEDBACK AMPLIFIER

This is a discrete circuit analysis method.

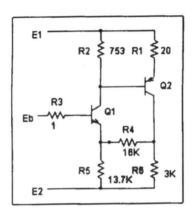

$K := 10^3$ $uA := 10^{-6}$ $pA := 10^{-12}$ $E1 := 15$

$R1 := 20$ $R2 := 753$ $R3 := 1$ $R4 := 16 \cdot K$

$R5 := 13.7 \cdot K$ $R6 := 3 \cdot K$ $E2 := -15$ $Eb := 5$

$B1 := 100$ $B2 := 100$ (Transistor betas)

Circuit equations:

$$\frac{E1 - V1}{R1} = (1 + B2) \cdot Ib2 \qquad \frac{E1 - V2}{R2} + Ib2 = B1 \cdot Ib1$$

$$\frac{Eb - V3}{R3} = Ib1 \qquad \frac{V4 - V5}{R4} + \frac{V4 - E2}{R5} = Ib1 \cdot (1 + B1)$$

$$\frac{V4 - V5}{R4} + \frac{V5 - E2}{R6} = B2 \cdot Ib2$$

$$Vbe_1 = V3 - V4 \qquad Vbe_2 = V1 - V2 \qquad \text{(7 equations)}$$

$$Vbe1 := 0.662 \qquad Vbe2 := 0.722$$

Start with guess $Vbe1 = Vbe2 = 0.6V$. Manually iterate two or three times with Vbe equations below.

$$A := \begin{bmatrix} \dfrac{1}{R1} & 0 & 0 & 0 & 0 & 0 & 1+B2 \\[2mm] 0 & \dfrac{1}{R2} & 0 & 0 & 0 & B1 & -1 \\[2mm] 0 & \dfrac{1}{R3} & 0 & 0 & 0 & 1 & 0 \\[2mm] 0 & 0 & 0 & \dfrac{1}{R4}+\dfrac{1}{R5} & \dfrac{-1}{R4} & -(1+B1) & 0 \\[2mm] 0 & 0 & 0 & \dfrac{-1}{R4} & \dfrac{1}{R4}+\dfrac{1}{R6} & 0 & -B2 \\[2mm] 0 & 0 & 1 & -1 & 0 & 0 & 0 \\[2mm] 1 & -1 & 0 & 0 & 0 & 0 & 0 \end{bmatrix} \qquad B := \begin{bmatrix} \dfrac{E1}{R1} \\[2mm] \dfrac{E1}{R2} \\[2mm] \dfrac{Eb}{R3} \\[2mm] \dfrac{E2}{R5} \\[2mm] \dfrac{E2}{R6} \\[2mm] Vbe1 \\[2mm] Vbe2 \end{bmatrix}$$

$$C := Isolve(A, B) \qquad Vo := C_5 \qquad Ib1 := C_6 \qquad Ib2 := C_7$$

$$C^T = [14.849 \quad 14.127 \quad 5 \quad 4.338 \quad 6.971 \quad 1.235 \cdot 10^{-5} \quad 7.488 \cdot 10^{-5}]$$

Vbe equations:

$$U := 30 \qquad Is := 3 \cdot pA \qquad Vbe1 := \dfrac{\ln\left[1 + \dfrac{(1+B1)\cdot Ib1}{Is}\right]}{U}$$

$$Vbe2 := \dfrac{\ln\left[1 + \dfrac{(1+B2)\cdot Ib2}{Is}\right]}{U} \qquad Vbe1 = 0.662 \qquad Vbe2 = 0.722$$

Solutions:

$$V_o = 6.971 \qquad Ib1 = 12.346 \ \text{°uA} \qquad Ib2 = 74.884 \ \text{°uA}$$

ITERATIVE TRANSIENT ANALYSIS

The principle used here starts from $e_L = L \cdot \dfrac{di_L}{dt}$ and $i_C = C \cdot \dfrac{dv_C}{dt}$
(ELI the ICE man).

Once an expression for e_L or i_C is developed, $\Delta i_L \ \Delta v_C$ and Δt are substituted for the differentials, and the expressions multiplied by Δt. Then initial conditions are established (usually zero), and iterations begin with a suitably small size for Δt.

A step input to a simple RC circuit is used to illustrate the method.

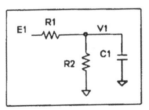

$$K := 10^3 \qquad uF := 10^{-6} \qquad\qquad us := 10^{-6} \qquad ms := 10^{-3}$$

$$R1 := 20 \cdot K \quad R2 := 40 \cdot K \qquad\qquad C1 := 0.1 \cdot uF$$

$$E1 := 10 \qquad dc := \frac{E1 \cdot R2}{R1 + R2} \qquad\qquad dc = 6.67$$

$$i_{C1} = i_1 - i_2 \qquad i_{C1} = \frac{E1 - V1}{R1} - \frac{V1}{R2} \qquad i_{C1} = \frac{E1}{R1} - V1 \cdot \left(\frac{1}{R1} + \frac{1}{R2} \right)$$

Substitute and multiply by Δt:

$$i_c = C1 \cdot \frac{\Delta V1}{\Delta t} \qquad C1 \cdot \Delta V1 = -V1 \cdot \left(\frac{1}{R1} + \frac{1}{R2}\right) \cdot \Delta t + \frac{E1 \cdot \Delta t}{R1}$$

$$T1 := \frac{1}{R1 \cdot C1} \qquad\qquad T2 := \frac{1}{C1} \cdot \left(\frac{1}{R1} + \frac{1}{R2}\right)$$

$$T1 = 500 \qquad\qquad T2 = 750 \qquad\qquad \frac{1}{T2} = 1.333 \circ ms$$

Choose $\Delta t <$ than $1/T2$:

$\Delta t := 100 \cdot us \quad$ Begin iteration: $V_1 := 0$

$\Delta Vc_1 := (E1 \cdot T1 - V_1 \cdot T2) \cdot \Delta t \quad \Delta Vc_1 = 0.5 \qquad V_2 := \Delta Vc_1 + V_1$

$\quad V_2 = 0.5$

$\Delta Vc_2 := (E1 \cdot T1 - V_2 \cdot T2) \cdot \Delta t \quad \Delta Vc_2 = 0.462 \qquad V_3 := \Delta Vc_2 + V_2$

$\quad V_3 = 0.962$

$\Delta Vc_3 := (E1 \cdot T1 - V_3 \cdot T2) \cdot \Delta t \quad \Delta Vc_3 = 0.428 \qquad V_4 := \Delta Vc_3 + V_3$

$\quad V_4 = 1.39 \qquad$ etc.

Program the sequence: $kmax := 100 \qquad k := 0..kmax$

$\qquad T := kmax \cdot \Delta t \qquad\qquad t := 0, \Delta t .. T$

$$DT := \begin{vmatrix} V_1 \leftarrow 0 \\ \text{for } k \in 1..k\,max \\ \qquad \begin{vmatrix} \Delta Vc_{k-1} \leftarrow (E1 \cdot T1 - V_{k-1} \cdot T2) \\ V_k \leftarrow \Delta Vc_{k-1} + V_{k-1} \end{vmatrix} \\ V \end{vmatrix}$$

Compare with inverse Laplace transform:

$$f(t) := \frac{E1 \cdot T1}{T2} \cdot (1 - \exp(-t \cdot T2))$$

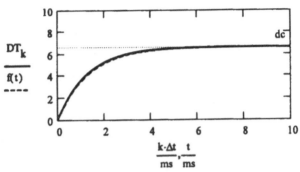

FIGURE 64. RC step response.

$$pk := 9.4973 \cdot 10^{-2} \qquad \text{Error:} \quad \frac{pk}{dc} = 1.42 \circ \%$$

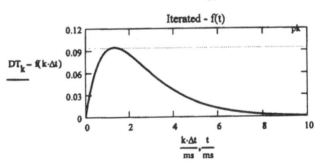

FIGURE 65. Iterated — f(t).

Hence the error using the iterative transient analysis method for this example is not excessive.

HALF-WAVE RECTIFIER TRANSIENT ANALYSIS

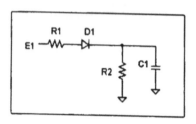

$$ms := 10^{-3} \qquad us := 10^{-6} \qquad pA := 10^{-12} \quad uA := 10^{-6}$$

$$F := 400 \qquad \omega := 2 \cdot \pi \cdot F \qquad T := \frac{1}{F}$$

$$Tn := 6 \quad \text{(No. of periods)} \qquad Vm := 10 \qquad \Delta t := 100 \cdot us$$

$$kmax := \frac{Tn \cdot T}{\Delta t} \qquad kmax = 150$$

$$t := 0, \Delta t \, .. \, Tn \cdot T \qquad u := 1 .. kmax \qquad K := 10^3$$

$$R1 := 20 \cdot K$$

$$R2 := 40 \cdot K \qquad uF := 10^{-6} \qquad C1 := 0.1 \cdot uF$$

$$E1(t) := Vm \cdot \sin(\omega \cdot t) \qquad T1 := \frac{1}{R1 \cdot C1} \qquad T2 := \frac{1}{C1} \cdot \left(\frac{1}{R1} + \frac{1}{R2} \right)$$

$$T3 := \frac{-1}{R2 \cdot C1}$$

Set R1 = infinity if output > input.

$$d := \begin{vmatrix} V_1 \leftarrow 0 \\ \text{for } k \in 2 .. k \max \\ \begin{vmatrix} \text{if } V_{k-1} < El(k \cdot \Delta t) \\ \begin{vmatrix} Tc \leftarrow T2 \\ \Delta V_{k-1} \leftarrow \left(T1 \cdot El(k \cdot \Delta t) + Tc \cdot V_{k-1}\right) \cdot \Delta t \end{vmatrix} \\ \text{otherwise} \\ \begin{vmatrix} Tc \leftarrow T3 \\ \Delta V_{k-1} \leftarrow Tc \cdot V_{k-1} \cdot \Delta t \end{vmatrix} \\ V_k \leftarrow \Delta V_{k-1} + V_{k-1} \end{vmatrix} \\ V \end{vmatrix}$$

Suggested refinements: insert diode voltage:

$$U := 30 \qquad Is := 3 \cdot pA \qquad Id := \frac{Vm}{R1 + R2} \qquad Id = 166.67 \circ uA$$

$$Vd := \frac{\ln\left(1 + \dfrac{Id}{Is}\right)}{U} \qquad Vd = 0.628 \qquad k\max \cdot \Delta t = 15 \circ ms$$

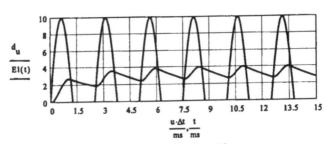

FIGURE 66. Halfwave rectifier.

SECOND-ORDER TRANSIENT ANALYSIS

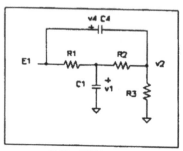

Bridged T-Network

$$K := 10^3 \qquad R1 := 100 \cdot K \qquad R2 := 80 \cdot K \qquad R3 := 200 \cdot K$$

$$uF := 10^{-6} \qquad C1 := 0.1 \cdot uF \qquad C4 := 0.2 \cdot uF \qquad us := 10^{-6}$$

$$G1 := \frac{1}{R1} + \frac{1}{R2} \qquad G2 := \frac{1}{R2} \qquad ms := 10^{-3} \qquad G4 := \frac{1}{R2} + \frac{1}{R3}$$

$$E1 := 1 \qquad u := E1 \qquad \text{(See derivation on page 170)}$$

Dc solution:

$$Q := -\begin{bmatrix} G1 & G2 \\ G2 & G4 \end{bmatrix} \qquad S := \begin{bmatrix} G1 \\ G4 \end{bmatrix} \qquad X := Q^{-1} \cdot S \cdot u$$

$$X = \begin{bmatrix} 7.368 \\ 4.737 \end{bmatrix}$$

$$P := -\begin{bmatrix} C1 & 0 \\ 0 & C4 \end{bmatrix} \qquad W := \begin{bmatrix} 1 & 0 \\ 0 & 1 \end{bmatrix}$$

$$A := (W \cdot P)^{-1} \cdot Q \qquad B := (W \cdot P)^{-1} \cdot S$$

Δt criteria:

$$Ta := \left| \frac{1}{max(A)} \right| \qquad Ta = 16 \circ ms \qquad \text{Use } \Delta t < Ta$$

$$\Delta t := 2 \cdot ms$$

$$kmax := 100 \qquad k := 2..kmax \qquad T := kmax \cdot \Delta t$$

$$T = 0.2$$

$$t := 0, \Delta t..T$$

Create input pulse: note: $\Phi(x)$ is the unit step function.

$$pulse(x, w) := \Phi(x) - \Phi(x - w) \qquad bp(x, f, w) := pulse(x - f, w)$$

$$E1(t) := bp(t, 0.05 \cdot T, 0.4 \cdot T)$$

Use matrix equations: Sometimes referred to as "seeded iteration."

$$\begin{bmatrix} \Delta V1_1 \\ \Delta V4_1 \end{bmatrix} := B \cdot E1(\Delta t) \cdot \Delta t$$

$$\begin{bmatrix} \Delta V1_k \\ \Delta V4_k \end{bmatrix} := A \cdot \begin{bmatrix} \Delta V1_{k-1} \\ \Delta V4_{k-1} \end{bmatrix} \cdot \Delta t + B \cdot E1(k \cdot \Delta t) \cdot \Delta t + \begin{bmatrix} \Delta V1_{k-1} \\ \Delta V4_{k-1} \end{bmatrix}$$

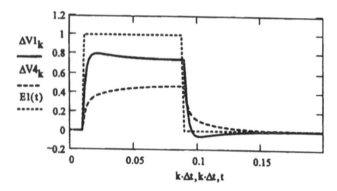

FIGURE 67.   Input and capacitor waveforms.

$$V2_k := E1(k \cdot \Delta t) - \Delta V4_k$$

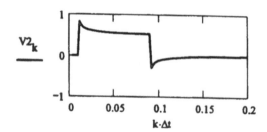

FIGURE 68   Output waveform.

DERIVATION OF MATRICES

Rule 1.   Use the following equation as required:

$$e_L = L \cdot \frac{di_L}{dt} \qquad \text{or} \qquad i_C = C \cdot \frac{dv_C}{dt}$$

Rule 2. To avoid excessive algebra, express the initial KCL or KVL equations in terms of i_C, i_L, v_C, or e_L.

Rule 3. Place e_L and i_C on left-hand side; i_L and v_C on right-hand side. When all terms include only these unknowns, the equation is complete and coefficients can be inserted into the matrices.

$$i_{c1} = i_1 - i_2 \quad i_{c1} = \frac{E1 - V1}{R1} - \frac{V1 - V2}{R2}$$

$$i_{c1} = \frac{E1}{R1} - V1 \cdot \left(\frac{1}{R1} + \frac{1}{R2}\right) + \frac{V2}{R2}$$

Eliminate V2:

$$V2 = E1 - V4 \qquad i_{c1} = E1 \cdot \left(\frac{1}{R1} + \frac{1}{R2}\right) - V1 \cdot \left(\frac{1}{R1} + \frac{1}{R2}\right) - \frac{V4}{R2}$$

$$G1 := \frac{1}{R1} + \frac{1}{R2} \qquad G2 := \frac{1}{R2} \quad i_{c1} = (E1 \cdot G1 - V1 \cdot G1 - V4 \cdot G2)$$

X-complete

$$i_{c4} + i_2 = i_3$$

$$i_{c4} = i_3 - i_2 = \frac{V2}{R3} - \frac{V1 - V2}{R2} = \frac{-V1}{R2} + V2 \cdot \left(\frac{1}{R2} + \frac{1}{R3}\right)$$

$$V2 = E1 - V4$$

$$i_{c4} = E1 \cdot \left(\frac{1}{R2} + \frac{1}{R3}\right) - \frac{V1}{R2} - V4 \cdot \left(\frac{1}{R2} + \frac{1}{R3}\right)$$

$$G4 := \frac{1}{R2} + \frac{1}{R3}$$

$$i_{C4} = (E1 \cdot G4 - V1 \cdot G2 - V4 \cdot G4) \cdot \Delta t \qquad (X)$$

$$E1 := 10 \qquad u := E1 \qquad C1 \cdot \Delta V1 = i_{C1} \cdot \Delta t \qquad C4 \cdot \Delta V4 = i_{C4} \cdot \Delta t$$

Matrix equations:

$$\begin{bmatrix} C1 \cdot \Delta V1 \\ C4 \cdot \Delta V4 \end{bmatrix} = Q \cdot \begin{bmatrix} V1 \\ V4 \end{bmatrix} \cdot \Delta t + S \cdot E1 \cdot \Delta t$$

Equation form:

$$W \cdot P \cdot \Delta x = Q \cdot x \cdot \Delta t + S \cdot u \cdot \Delta t$$

where

$$W := \begin{bmatrix} 1 & 0 \\ 0 & 1 \end{bmatrix} \qquad P := \begin{bmatrix} C1 & 0 \\ 0 & C4 \end{bmatrix} \qquad Q := \begin{bmatrix} G1 & G2 \\ G2 & G4 \end{bmatrix}$$

$$S := \begin{bmatrix} G1 \\ G4 \end{bmatrix} \qquad x = \begin{bmatrix} V1 \\ V4 \end{bmatrix} \qquad \Delta x = \begin{bmatrix} \Delta V1 \\ \Delta V4 \end{bmatrix}$$

To get state space standard form:

$$\Delta x = A \cdot x \cdot \Delta t + B \cdot u \cdot \Delta t \qquad \text{pre-multiply by } (W \cdot P)^{-1}$$

$$\Delta x = (W \cdot P)^{-1} \cdot Q \cdot x \cdot \Delta t + (W \cdot P)^{-1} \cdot S \cdot u \cdot \Delta t$$

Then:

$$A := (W \cdot P)^{-1} \cdot Q \qquad B := (W \cdot P)^{-1} \cdot S$$

First iteration:

$$x = 0; \; E1 = E1(t) = E1(\Delta t) \qquad \begin{bmatrix} \Delta V1_1 \\ \Delta V4_1 \end{bmatrix} = S \cdot E1(\Delta t) \cdot \Delta t$$

Second and subsequent iterations: $k = 2, 3, \ldots$

$$\begin{bmatrix} \Delta V1_k \\ \Delta V4_k \end{bmatrix} = A \cdot \begin{bmatrix} \Delta V1_{k-1} \\ \Delta V4_{k-1} \end{bmatrix} \cdot \Delta t + B \cdot E1\,(k \cdot \Delta t) \cdot \Delta t + \begin{bmatrix} \Delta V1_{k-1} \\ \Delta V4_{k-1} \end{bmatrix}$$

DC solution:

$$\text{Set } \frac{\Delta x}{\Delta t} = 0 \qquad Q \cdot X + S \cdot u = 0 \qquad X := -Q{-1} \cdot S \cdot u \qquad X = \begin{bmatrix} 7.368 \\ 4.737 \end{bmatrix}$$

Using the method given above, a considerable amount of algebraic manipulation is avoided in obtaining the A and B matrices. (See Reference 14.) The labor savings is not realized in this simple second order circuit. However in the sixth-order pulse transformer analysis following, the considerable task of isolating each state variable on the left-hand side is not necessary — the computer does the job for you.

TRANSIENT MCA OF BRIDGED-T CIRCUIT

Transient MCA is performed in the same manner as in ac and dc circuits.

$$Tr := 0.02 \qquad Tc := 0.1 \qquad Nc := 5 \qquad N := 1000 \qquad m := 1..N$$

$$w := 1..Nc \qquad T := \begin{bmatrix} -Tr & -Tr & -Tr & -Tc & -Tc \\ Tr & Tr & Tr & Tc & Tc \end{bmatrix}$$

$$Tn_{m,w} := (T_{2,w} - T_{1,w}) \cdot rnd(1) + T_{1,w} + 1$$

$$Qr_m := -1 \cdot \begin{bmatrix} \dfrac{1}{R1 \cdot Tn_{m,1}} + \dfrac{1}{R2 \cdot Tn_{m,2}} & \dfrac{1}{R2 \cdot Tn_{m,2}} \\[3mm] \dfrac{1}{R2 \cdot Tn_{m,2}} & \dfrac{1}{R2 \cdot Tn_{m,1}} + \dfrac{1}{R3 \cdot Tn_{m,3}} \end{bmatrix}$$

$$Sr_m := \begin{bmatrix} \dfrac{1}{R1 \cdot Tn_{m,1}} + \dfrac{1}{R2 \cdot Tn_{m,2}} \\ \dfrac{1}{R2 \cdot Tn_{m,1}} + \dfrac{1}{R3 \cdot Tn_{m,3}} \end{bmatrix}$$

$$Pr_m := \begin{bmatrix} C1 \cdot Tn_{m,4} & 0 \\ 0 & C4 \cdot Tn_{m,5} \end{bmatrix}$$

$$Ar_m := (W \cdot Pr_m)^{-1} \cdot Qr_m \qquad Br_m := (W \cdot Pr_m)^{-1} \cdot Sr_m$$

$$\begin{bmatrix} \Delta V1_{m,1} \\ \Delta V4_{m,1} \end{bmatrix} := Br_m \cdot El(\Delta t) \cdot \Delta t$$

$$\begin{bmatrix} \Delta V1_{m,k} \\ \Delta V4_{m,k} \end{bmatrix} := Ar_m \cdot \begin{bmatrix} \Delta V1_{m,k-1} \\ \Delta V4_{m,k-1} \end{bmatrix} \cdot \Delta t + Br_m \cdot El(k \cdot \Delta t) \cdot \Delta t \\ + \begin{bmatrix} \Delta V1_{m,k-1} \\ \Delta V4_{m,k-1} \end{bmatrix}$$

$$V2r_{m,k} := El(k \cdot \Delta t) - \Delta V4_{m,k}$$

$$V2h_k := if(max(V2r^{<k>}) > 0, \ max(V2r^{<k>}), \ min(V2r^{<k>}))$$

(The if statements are necessary because of the bipolar output.)

$$V2l_k := if(min(V2r^{<k>}) > 0, \ min(V2r^{<k>}), \ max(V2r^{<k>}))$$

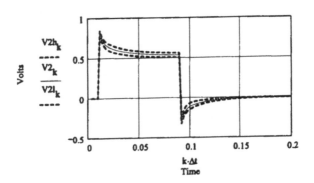

FIGURE 69. MCA of transient response. (N = 1000).

PULSE TRANSFORMER – FREQUENCY RESPONSE

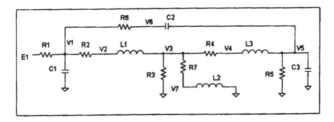

Transformer Model.

$R1 := 10$ $K := 10^3$ $R2 := 1.5$ $R3 := 20 \cdot K$

$R4 := 1.5$ $R5 := 1 \cdot K$ $R6 := 0.5$

$pF := 10^{-12}$ $C1 := 20 \cdot pF$ $C2 := 5 \cdot pF$ $C3 := 20 \cdot pF$

$uH := 10^{-6}$ $mH := 10^{-3}$

$L1 := 1 \cdot uH \qquad L2 := 2 \cdot mH \qquad L3 := 1 \cdot uH \qquad uA := 10^{-6}$

$R7 := 1 \qquad\qquad ns := 10^{-9} \qquad MHz := 10^{6}$

$BF := 2 \qquad\qquad ND := 6 \qquad\qquad PD := 20 \qquad\qquad P := 2 \cdot \pi$

$i := 1 .. ND \cdot PD + 1 \qquad\qquad L_i := BF + \dfrac{i-1}{PD}$

$F_i := 10^{L_i} \qquad\qquad s_i := P \cdot F_i \cdot \sqrt{-1} \qquad Z2_i := R2 + s_i \cdot L1$

$Z6_i := R6 + \dfrac{1}{s_i \cdot C2} \qquad Z4_i := R4 + s_i \cdot L3$

$Z5_i := \dfrac{R5}{s_i \cdot R5 \cdot C3 + 1} \qquad Z7_i := \dfrac{R3 \cdot (R7 + s_i \cdot L2)}{R3 + R7 + s_i \cdot L2}$

$$A_i := \begin{bmatrix} \dfrac{1}{R1} + s_i \cdot C1 + \dfrac{1}{Z2_i} + \dfrac{1}{Z6_i} & \dfrac{-1}{Z2_i} & 0 \\[2mm] \dfrac{-1}{Z2_i} & \dfrac{1}{Z2_i} + \dfrac{1}{Z7_i} + \dfrac{1}{Z4_i} & \dfrac{-1}{Z4_i} \\[2mm] 0 & \dfrac{-1}{Z4_i} & \dfrac{1}{Z6_i} + \dfrac{1}{Z5_i} + \dfrac{1}{Z4_i} \end{bmatrix}$$

$$B := \begin{bmatrix} \dfrac{1}{R1} \\[2mm] 0 \\[1mm] 0 \end{bmatrix} \qquad C_i := \text{lsolve}(A_i, B)$$

$Vo_i := 20 \cdot \log \left[\left| (C_i)_3 \right| \right]$

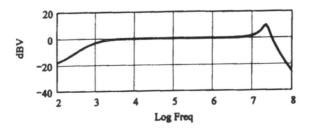

FIGURE 70. Frequency response.

Flat response from about 1 KHz to 10 MHz.

PULSE TRANSFORMER — TRANSIENT RESPONSE

See page 175 for a schematic and component values. Create required matrices (Note that dx and x are for bookkeeping purposes only. See derivation on page 182):

$$
dx = \begin{bmatrix} e_{L1} \\ e_{L2} \\ e_{L3} \\ i_{C1} \\ i_{C2} \\ i_{C3} \end{bmatrix} \quad
W := \begin{bmatrix} 0 & 0 & 0 & 1 & 1 & 0 \\ \dfrac{1}{R2} & \dfrac{1}{R2} & 0 & 0 & 0 & 0 \\ 0 & \dfrac{1}{R3} & 0 & 0 & 0 & 0 \\ 0 & \dfrac{1}{R4} & \dfrac{-1}{R4} & 0 & 0 & 0 \\ 0 & 0 & 0 & 0 & 1 & -1 \\ 0 & 0 & 0 & 0 & 1 & 0 \end{bmatrix} \quad
P := \text{diag} \begin{bmatrix} L1 \\ L2 \\ L3 \\ C1 \\ C2 \\ C3 \end{bmatrix}
$$

$$Q := \begin{bmatrix} -1 & 0 & 0 & \dfrac{-1}{R1} & 0 & 0 \\ -1 & \dfrac{-R7}{R2} & 0 & \dfrac{1}{R2} & 0 & 0 \\ 1 & -\left(1 + \dfrac{R7}{R3}\right) & -1 & 0 & 0 & 0 \\ 0 & \dfrac{-R7}{R4} & 1 & 0 & 0 & \dfrac{1}{R4} \\ 0 & 0 & -1 & 0 & 0 & \dfrac{1}{R5} \\ 0 & 0 & 0 & \dfrac{1}{R6} & \dfrac{-1}{R6} & \dfrac{-1}{R6} \end{bmatrix} \quad x = \begin{bmatrix} i_{L1} \\ i_{L2} \\ i_{L3} \\ V1 \\ Vc2 \\ V5 \end{bmatrix} \quad S := \begin{bmatrix} \dfrac{1}{R1} \\ 0 \\ 0 \\ 0 \\ 0 \\ 0 \end{bmatrix} \quad \begin{array}{l} E1 := 10 \\ u := E1 \end{array}$$

DC solution:

$$X := -Q^{-1} \cdot S \cdot u \qquad X = \begin{bmatrix} 0.800 \\ 0.799 \\ 7.980 \cdot 10^{-4} \\ 1.999 \\ 1.201 \\ 0.798 \end{bmatrix}$$

Example: $i_{L1} = 0.8$ Adc; $V5 = 0.798$ Vdc.

$$A := (W \cdot P)^{-1} \cdot Q \qquad B := (W \cdot P)^{-1} \cdot S$$

$$Ta := \left| \frac{1}{\max(A)} \right| \qquad Tb := \left| \frac{1}{\min(A)} \right|$$

$Ta = 2.5$ °ps $Tb = 2.5$ °ps $\Delta t := 2 \cdot ps$ $kmax := 1 \cdot 10^5$

$T := kmax \cdot \Delta t$ $T = 200$ °ns

Create trapezoidal input:

$$\text{ramp}(p, x, a) := p \cdot (x - a) \cdot \Phi(x - a) \qquad \text{tmax} := \text{kmax} \cdot \Delta t$$

$$\text{tmax} := 200 \cdot \text{ns} \qquad t := 0, \Delta t \ldots T$$

$$\text{trap}(p, x, a, b, c, d) := \text{ramp}(p, x, a) - \text{ramp}(p, x, b) - \text{ramp}(p, x, c)$$
$$+ \text{ramp}(p, x, d)$$

$$\text{rt} := 5 \cdot \Delta t$$

$$\text{trap}(p, x, a, b, c, d) := \text{ramp}(p, x, a) - \text{ramp}(p, x, b) - \text{ramp}(p, x, c)$$
$$+ \text{ramp}(p, x, d)$$

$$\text{rt} := 5 \cdot \Delta t$$

From page 177, the maximum frequency response is:

$$F := 10 \cdot \text{MHz}$$

Rise time vs. frequency:

$$Tr := \frac{\ln(9)}{2 \cdot \pi \cdot F} \qquad Tr = 34.97 \cdot \text{ns}$$

Use Δt less than the smaller of Ta and Tb. Note that the input pulse rise/fall time is 10ps which is much less than Tr. Hence we should expect the output to be distorted.

$$\text{pw} := 50000 \cdot \Delta t \qquad \text{dt} := \Delta t \qquad a := \text{dt} \qquad b := a + \text{rt} \qquad c := \text{pw} + b$$

$$d := c + \text{rt} \qquad p := \frac{10}{b - a} \qquad E1(t) := \text{trap}(p, t, a, b, c, d)$$

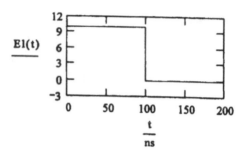

FIGURE 71. Input pulse.

$\Delta t = 2 \cdot ps \quad rt = 10 \cdot ps \quad T = 200 \cdot ns \quad k := 2 .. kmax \quad pw = 100 \cdot ns$

$$\begin{bmatrix} \Delta iL1_1 \\ \Delta iL2_1 \\ \Delta iL3_1 \\ \Delta V1_1 \\ \Delta Vc2_1 \\ \Delta V5_1 \end{bmatrix} := B \cdot El(\Delta t) \cdot \Delta t$$

$$\begin{bmatrix} \Delta iL1_k \\ \Delta iL2_k \\ \Delta iL3_k \\ \Delta V1_k \\ \Delta Vc2_k \\ \Delta V5_k \end{bmatrix} := A \cdot \begin{bmatrix} \Delta iL1_{k-1} \\ \Delta iL2_{k-1} \\ \Delta iL3_{k-1} \\ \Delta V1_{k-1} \\ \Delta Vc2_{k-1} \\ \Delta V5_{k-1} \end{bmatrix} \cdot \Delta t + B \cdot El(k \cdot \Delta t) \cdot \Delta t + \begin{bmatrix} \Delta iL1_{k-1} \\ \Delta iL2_{k-1} \\ \Delta iL3_{k-1} \\ \Delta V1_{k-1} \\ \Delta Vc2_{k-1} \\ \Delta V5_{k-1} \end{bmatrix}$$

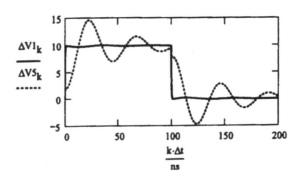

FIGURE 72. C1 and C3 voltage.

$$pw = 100 \circ ns \qquad rt = 10 \circ ps \qquad \Delta t = 2 \circ ps \qquad T = 200 \circ ns$$

Note the distortion (ringing) on the output V5.

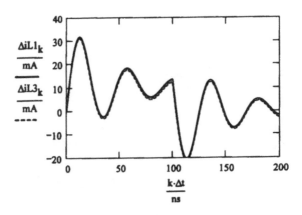

FIGURE 73. L1 and L3 current.

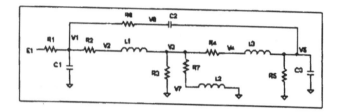

Derivation: As a rule e_L and i_C are on the left-hand side; i_L and v_C are on the right-hand side. As much as possible, express the initial KCL or KVL equations in terms of i_C, i_L, v_C, or e_L. For example, do not write

$$i_{R1} = i_{C1} + i_{R2} + i_{C2}$$

for node V1 below; use i_{L1} instead of i_{R2}.

Using this method, much algebra is avoided and the required equations are easily derived.

Node V1:

$$i_{R1} = i_{C1} + i_{L1} + i_{C2} \qquad i_{C1} + i_{C2} = i_{R1} - i_{L1}$$

$$i_{C1} + i_{C2} = \frac{E1}{R1} - \frac{V1}{R1} - i_{L1} \qquad (X)$$

Node V2:

$$\frac{V1}{R2} - \frac{V2}{R2} - i_{L1} = 0 \qquad V2 = e_{L1} + V3 \qquad V3 = i_{L2} \cdot R7 + e_{L2}$$

$$V2 = e_{L1} + i_{L2} \cdot R7 + e_{L2} \qquad \frac{V1}{R2} - \frac{e_{L1}}{R2} - \frac{i_{L2} \cdot R7}{R2} - \frac{e_{L2}}{R2} = i_{L1}$$

$$\frac{e_{L1}}{R2} + \frac{e_{L2}}{R2} = \frac{V1}{R2} - \frac{i_{L2} \cdot R7}{R2} - i_{L1} \qquad (X)$$

Node V3:

$$i_{L1} = i_{R3} + i_{L3} + i_{L2} \quad i_{L1} - i_{L2} - i_{L3} = i_{R3} \quad i_{L1} - i_{L2} - i_{L3} = \frac{V3}{R3}$$

$$V3 = i_{L2} \cdot R7 + e_{L2} \quad i_{L1} - i_{L2} - i_{L3} = \frac{i_{L2} \cdot R7}{R3} + \frac{e_{L2}}{R3}$$

$$\frac{e_{L2}}{R3} = i_{L1} - i_{L2} \cdot \left(1 + \frac{R7}{R3}\right) - i_{L3} \tag{X}$$

Node V4:

$$\frac{V3}{R4} - \frac{V4}{R4} - i_{L3} = 0 \quad V4 = e_{L3} + V5 \quad V3 = i_{L2} \cdot R7 + e_{L2}$$

$$\frac{i_{L2} \cdot R7}{R4} + \frac{e_{L2}}{R4} - \frac{e_{L3}}{R4} - \frac{V5}{R4} - i_{L3} = 0$$

$$\frac{e_{L2}}{R4} - \frac{e_{L3}}{R4} = \frac{i_{L2} \cdot R7}{R4} + \frac{V5}{R4} + i_{L3} \tag{X}$$

Node V5:

$$i_{L3} + i_{C2} = i_{R5} + i_{C3} \quad i_{C2} - i_{C3} = i_{R5} - i_{L3} \quad i_{C2} - i_{C3} = \frac{V5}{R5} - i_{L3} \tag{X}$$

Node V6:

$$i_{R6} - i_{C2} = 0 \quad \frac{V1 - V6}{R6} - i_{C2} = 0 \quad V6 = Vc2 + V5$$

$$\frac{V1}{R6} - \frac{V6}{R6} - i_{C2} = 0 \quad \frac{V1}{R6} - \frac{Vc2}{R6} - \frac{V5}{R6} - i_{C2} = 0$$

$$i_{C2} = \frac{V1}{R6} - \frac{Vc2}{R6} - \frac{V5}{R6} \tag{X}$$

SPICE LISTING

To show that the foregoing transient analysis methods are correct, the reader is encouraged to run the following Spice netlist and compare with the plots given on page 181.

```
TRANSFORMER PULSE RESPONSE
* File: xformer2.cir
VIN 1 0 PWL(0,0 2ps,0 12ps,10 100.012ns,10
    10 100.022ns,0)
* 10ps rise & fall time; 100ns pulse width
R1 1 2 10
C1 2 0 20pF
R2 2 3 1.5
R6 2 7 0.5
L1 3 4 1uH
R3 4 0 20K
R7 4 8 1
L2 8 0 2mH
R4 4 5 1.5
L3 5 6 1uH
R5 6 0 1K
C3 6 0 20pF
C2 7 6 5pF
.TRAN 1ns, 200ns
.PROBE V(2) V(6) I(L1) I(L3)
.OPTIONS ITL5=0
.OPTIONS NOECHO NOPAGE NOMOD
.END
```

Appendix

DERIVATION OF THE RSS EQUATION:

Using three resistors as an example, the RSS equation is:

$$\left(\sigma_O\right)^2 = \left(\frac{\partial Vo}{\partial R1}\cdot\sigma_{R1}\right)^2 + \left(\frac{\partial Vo}{\partial R2}\cdot\sigma_{R2}\right)^2 + \left(\frac{\partial Vo}{\partial R3}\cdot\sigma_{R3}\right)^2$$

This definition is derived from the total differential in calculus which, for the function $Vo = f(R1, R2, R3)$ is:

$$dVo = \frac{\partial Vo}{\partial R1}\cdot dR1 + \frac{\partial Vo}{\partial R2}\cdot dR2 + \frac{\partial Vo}{\partial R3}\cdot dR3$$

The dR's are differentials and thus are infinitesimally small and approach zero. When the dR's are set to the component tolerances, this is stretching the definition somewhat. Thus with large component tolerances (and sensitivities) *RSS becomes a poor estimator and should not be used.*

The RSS method is also formally defined as the variance of a non-linear function in Reference 7.

$$\mathrm{Var}\left[\,g\left(X_1, X_2, ..., X_n\right)\right] \approx \sum_{i=1}^{n}\sum_{j=1}^{n}\left(\frac{\partial g}{\partial X_i}\right)\cdot\left(\frac{\partial g}{\partial X_j}\right)\cdot\mathrm{Cov}\left[X_i, X_j\right]$$

where $\mathrm{Cov}[X_i, X_j]$ is the covariance of X_i and X_j. If the variables X_i are statistically independent, then the right-hand side becomes

$$\sum_{i=1}^{n}\left(\frac{\partial g}{\partial X_i}\right)^2 \cdot \mathrm{Var}\left[X_i\right]$$

For the derivation given in Reference 8, let U be some function of the three measured quantities X, Y, and Z. That is

$$U = f(X, Y, Z)$$

If each independent variable is allowed to change by a small amount dX, dY, dZ, then the quantity U will change by an amount dU given by the total differential

$$dU = \frac{\partial U}{\partial X} \cdot dX + \frac{\partial U}{\partial Y} \cdot dY + \frac{\partial U}{\partial Z} \cdot dZ$$

Applying this to a set of measurements, let $\Delta x = dX$, $\Delta y = dY$, and $\Delta z = dZ$, be a set of residuals of the measured quantities. Then letting $\Delta u = dU$

$$\Delta u_1 = \frac{\partial U}{\partial X} \Delta x_1 + \frac{\partial U}{\partial Y} \Delta y_1 + \frac{\partial U}{\partial Z} \Delta z_1$$

$$\Delta u_2 = \frac{\partial U}{\partial X} \Delta x_2 + \frac{\partial U}{\partial Y} \Delta y_2 + \frac{\partial U}{\partial Z} \Delta z_2$$

$$\bullet$$
$$\bullet$$
$$\bullet$$

$$\Delta u_n = \frac{\partial U}{\partial X} \Delta x_n + \frac{\partial U}{\partial Y} \Delta y_n + \frac{\partial U}{\partial Z} \Delta z_n$$

Squaring both sides and adding gives

$$\Delta u_1^2 = \left(\frac{\partial U}{\partial X}\right)^2 \Delta x_1^2 + 2\left(\frac{\partial U}{\partial X}\right)\left(\frac{\partial U}{\partial Y}\right)\Delta x_1 \, \Delta y_1 + \ldots + \left(\frac{\partial U}{\partial Z}\right)^2 \Delta z_1^2$$

$$\Delta u_2^2 = \left(\frac{\partial U}{\partial X}\right)^2 \Delta x_2^2 + 2\left(\frac{\partial U}{\partial X}\right)\left(\frac{\partial U}{\partial Y}\right)\Delta x_2\, \Delta y_2 + ... + \left(\frac{\partial U}{\partial Z}\right)^2 \Delta z_2^2$$

$$\vdots$$

$$\Delta u_n^2 = \left(\frac{\partial U}{\partial X}\right)^2 \Delta x_n^2 + 2\left(\frac{\partial U}{\partial X}\right)\left(\frac{\partial U}{\partial Y}\right)\Delta x_n\, \Delta y_n + ... + \left(\frac{\partial U}{\partial Z}\right)^2 \Delta z_n^2$$

or

$$\sum \Delta u^2 = \left(\frac{\partial U}{\partial X}\right)^2 \sum \Delta x^2 + 2\left(\frac{\partial U}{\partial X}\right)\left(\frac{\partial U}{\partial Y}\right)$$

$$\sum \Delta x\, \Delta y + ... + \left(\frac{\partial U}{\partial Z}\right)^2 \sum \Delta z^2$$

If the measured quantities are statistically independent, i.e., uncorrelated, the cross products are zero. Dropping these cross products and dividing both sides by $n - 1$ gives:

$$\frac{\sum \Delta u^2}{n-1} = \left(\frac{\partial U}{\partial X}\right)^2 \frac{\sum \Delta x^2}{n-1} + \left(\frac{\partial U}{\partial Y}\right)^2 \frac{\sum \Delta y^2}{n-1} + \left(\frac{\partial U}{\partial Z}\right)^2 \frac{\sum \Delta z^2}{n-1}$$

or finally,

$$\sigma_u^2 = \left(\frac{\partial U}{\partial X}\right)^2 \sigma_x^2 + \left(\frac{\partial U}{\partial Y}\right)^2 \sigma_y^2 + \left(\frac{\partial U}{\partial Z}\right)^2 \sigma_z^2$$

DERIVATION OF $\sqrt{3}$ FACTOR FOR UNIFORM DISTRIBUTION INPUTS

Repeat formal RSS definition for three resistors:

$$\text{var}(Vo) = \left(\frac{dVo}{dR1}\right)^2 \cdot \text{var}(R1) + \left(\frac{dVo}{dR2}\right)^2 \cdot \text{var}(R2)$$

$$+ \left(\frac{dVo}{dR3}\right)^2 \cdot \text{var}(R3)]$$

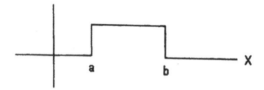

The variance of a uniform or rectangular distribution is defined as:

$$\text{var}(X) = \frac{(b-a)^2}{12} \qquad \sigma(X) = \frac{b-a}{\sqrt{12}}$$

For example, 1K, 1% resistor:

$$\sigma(R1) = \frac{1010 - 990}{\sqrt{12}} = \frac{2 \cdot R1 \cdot T1}{\sqrt{12}}$$

Insert var(X) into the first term and use approximations:

$$\left(\frac{\Delta Vo}{\Delta R1}\right)^2 \cdot \frac{(2 \cdot R1 \cdot T1)^2}{12} = \left(\frac{2 \cdot \Delta Vo \cdot R1 \cdot T1}{\Delta R1 \cdot \sqrt{12}}\right)^2 = \left(\frac{2 \cdot \Delta Vo \cdot T1}{dpf \cdot \sqrt{12}}\right)^2$$

$$= \left(\frac{\Delta Vo \cdot T1}{dpf \cdot \sqrt{3}}\right)^2 = \left(\frac{\sqrt{3} \cdot \Delta Vo \cdot T1}{3 \cdot dpf}\right)^2$$

$$\Delta Vrss = \frac{\sqrt{3 \cdot \left[(\Delta Vo1 \cdot T1)^2 + (\Delta Vo2 \cdot T2)^2 + (\Delta Vo3 \cdot T3)^2\right]}}{dpf}$$

$$\Delta Vrss = Vo \cdot \sqrt{3 \cdot \left[(Sen1 \cdot T1)^2 + (Sen2 \cdot T2)^2 + (Sen3 \cdot T3)^2\right]}$$

Example:

$$T1 := 0.01 \qquad R1 = 1000 \qquad 2 \cdot R1 \cdot T1 = 20$$

$$\sigma u := \frac{2 \cdot R1 \cdot T1}{\sqrt{12}} \qquad \sigma u = 5.774 \qquad 7\sqrt{3} \cdot \sigma u = 10$$

ASYMMETRIC GAUSSIAN DISTRIBUTION –
Spice .DISTRIBUTION Statement

For hopelessly addicted Spice users, the following will provide a means to calculate asymmetric (or symmetric) Gaussian distribution histogram values for the .MC .DISTRIBUTION option.

USER INPUT

$Rnom := 1000 \quad T1 := -0.1 \quad T2 := 0.02$ (1K resistor, +2%, −10%)

MATHCAD CALCULATIONS

$$N := 500 \qquad\qquad F1 := \text{if}(|T1| > |T2|, 0, 1)$$

$$\text{Rmin} := \text{Rnom} \cdot (1 + T1) \qquad \text{Rlow} := \text{Rnom} \cdot (1 - T2)$$

$$\text{Rmax} := \text{Rnom} \cdot (1 + T2) \qquad \text{Rplus} := \text{Rnom} \cdot (1 - T1)$$

$$nb := 20 \qquad inc := \frac{2}{nb} \qquad x := -1, -1 + inc .. 1$$

$$m1 := \frac{R\,max + R\,min}{2} \qquad\qquad s1 := \frac{R\,max - R\,min}{6}$$

$$\text{Rplus} := \text{if}(F1 = 0, \text{Rplus}, \text{Rmax}) \qquad \text{Rmin} := \text{if}(F1 = 0, \text{Rmin}, \text{Rlow})$$

$$intv := \frac{\text{Rplus} - R\,min}{nb}$$

$$y(r) := N \cdot intv \cdot dnorm(r, m1, s1) \qquad u(r) := \text{floor}(y(r) + 0.5)$$

$$r := \text{Rmin}, \text{Rmin} + intv .. \text{Rplus}$$

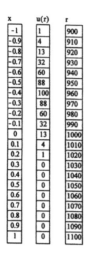

x	u(r)	r
-1	1	900
-0.9	4	910
-0.8	13	920
-0.7	32	930
-0.6	60	940
-0.5	88	950
-0.4	100	960
-0.3	88	970
-0.2	60	980
-0.1	32	990
0	13	1000
0.1	4	1010
0.2	1	1020
0.3	0	1030
0.4	0	1040
0.5	0	1050
0.6	0	1060
0.7	0	1070
0.8	0	1080
0.9	0	1090
1	0	1100

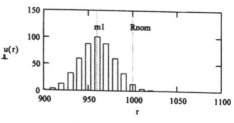

FIGURE 74. Histogram.

Using this example and naming the distribution ASYM,
the .DISTRIBUTION statement should be constructed from the x
and u(r) columns as follows:

.DISTRIBUTION ASYM (−1, 1), (−.9, 4) (−.8, 13) (−.7, 32)

$$(−.6, 60) (−.5, 88) (−.4, 100) (−.3, 88)$$

$$+ (−.2, 60) (−.1, 32) (0, 13) (.1, 4) (.2, 1) (.3, 0) (.4, 0) (.5, 0)$$

$$(.6, 0) (.7, 0) (.8, 0) (.9, 0) (1, 0)$$

Used for a resistor model named RA the .MODEL statement
should be:

.MODEL RA RES (R = 1 DEV/ASYM = 10%)
and similarly for inductors L and capacitors C.

RATIO ALGORITHMS

The author has found the following routines to be useful in analog
circuit design.

Step 1. (or Algorithm #1) Find best parallel resistor values:

Rp = R1//R2: User input: Rp := 55.86 Tol := 0.5

(Choose Tol = 2 for 2% values, 1 for 1% values, or 0.5 for
0.5% and 0.1% values. Applies to all four algorithms
below.)

Required functions:

$$\text{frac}(x) := x - \text{floor}(x) \qquad \text{rti}(x) := \text{floor}(x + 0.5)$$

$$B := \frac{96}{\text{Tol}} \qquad Lx(x) := \text{rti}(B \cdot \log(x))$$

$$Rx(x) := 10^{\text{floor}\left(\frac{x}{B}\right)} \cdot 0.01 \cdot \text{rti}\left(100 \cdot 10^{\text{frac}\left(\frac{x}{B}\right)}\right)$$

Trial values of R1:

$$m := 1 \ldots B + 1 \qquad R1_m := Rx(Lx(Rp) + m)$$

With these trial values of R1, solve for exact values R2t, convert to standard values R2; find min error.

$$R2t_m := \frac{Rp \cdot R1_m}{R1_m - Rp} \qquad R2_m := Rx(Lx(R2t_m))$$

$$E_m := |R2_m - R2t_m| \qquad Emin := \min(E)$$

$$ind := \begin{vmatrix} q \leftarrow 1 \\ \text{while } E_q > E \min \\ \quad q \leftarrow q + 1 \\ q \end{vmatrix}$$

$$ind = 99 \quad R1_{ind} = 182 \quad R2_{ind} = 80.6$$

Use $R1_{ind}$ and $R2_{ind}$ with appropriate decade values.

$$Rc_{ind} := \frac{R1_{ind} \cdot R2_{ind}}{R1_{ind} + R2_{ind}} \qquad Rcind = 55.861 \quad Rp = 55.86$$

$$pce := \frac{Rc_{ind}}{Rp} - 1 \qquad pce = 0.002 \circ\% \quad \textit{Percent error}$$

Step 2. (or Algorithm #2) Find best non-inverting gain values:

$$G = 1 + \frac{R1}{R2} > 1 \qquad \text{User input: } G := 2.88$$

$$m := 1 \ldots B + 1 \quad R1_m := Rx(Lx(100) + m) \quad R2t_m := \frac{R1_m}{G - 1}$$

$$R2_m := Rx(Lx(R2t_m)) \quad E_m := |R2_m - R2t_m| \quad Emin := \min(E)$$

$$ind := \begin{vmatrix} q \leftarrow 1 \\ \text{while } E_q > E\min \\ \quad q \leftarrow q + 1 \\ q \end{vmatrix}$$

$$Ind = 6 \qquad R1_{ind} = 107 \qquad R2_{ind} = 56.9$$

$$Gc_{ind} := 1 + \frac{R1_{ind}}{R2_{ind}} \qquad Gc_{ind} = 2.88 \qquad G = 2.88$$

$$pce := \frac{Gc_{ind}}{G} - 1 \qquad pce = 0.017 \circ \%$$

Step 3. (or Algorithm #3) Find best voltage divider values:

$$0 < D = \frac{R2}{R1 + R2} < 1 \qquad \text{User input:} \quad D := 0.2124$$

$$m := 1 .. B + 1 \qquad R1_m := Rx(Lx(100) + m)$$

$$R2t_m := \frac{R1_m \cdot D}{1 - D} \qquad R2_m := Rx(Lx(R2t_m))$$

$$E_m := |R2_m - R2t_m| \qquad Emin := \min(E)$$

$$ind := \begin{vmatrix} q \leftarrow 1 \\ \text{while } E_q > E\min \\ \quad q \leftarrow q + 1 \\ q \end{vmatrix}$$

$$ind = 7 \quad R1_{ind} = 109 \qquad R2_{ind} = 29.4$$

$$Dc_{ind} := \frac{R2_{ind}}{R1_{ind} + R2_{ind}} \qquad Dc_{ind} = 0.212 \qquad D = 0.212$$

$$\text{pce} := \frac{Dc_{ind}}{D} - 1 \qquad \text{pce} = 0.013 \circ\%$$

Step 4. (or Algorithm #4) Find best inverting gain values:

$$G = \frac{-R1}{R2} < 0 \qquad \text{User input:} \quad G := -3.861$$

$$m := 1..B + 1 \qquad R1_m := Rx(Lx(100) + m)$$

$$R2t_m := \frac{R1_m}{-G} \qquad R2_m := Rx(Lx(R2t_m))$$

$$E_m := |R2_m - R2t_m| \qquad \text{Emin} := \min(E)$$

$$\text{ind} := \begin{vmatrix} q \leftarrow 1 \\ \text{while } E_q > E\min \\ \quad q \leftarrow q + 1 \\ q \end{vmatrix}$$

$$\text{ind} = 6 \qquad R1_{ind} = 107 \qquad R2_{ind} = 27.7$$

$$Gn_{ind} := \frac{-R1_{ind}}{R2_{ind}} \qquad Gn_{ind} = -3.863 \qquad G = -3.861$$

$$\text{pce} := \frac{Gn_{ind}}{G} - 1 \qquad \text{pce} = 0.047 \circ\%$$

Can also find standard values of resistors Rs from calculated values Rc:

$$Rc := 18263 \qquad Rs := Rx(Lx(Rc)) \qquad Rs = 18200$$

One aberration out of 192:

$$Rc := 920 \qquad Rs := Rx(Lx(Rc)) \qquad Rs = 919$$

(Standard 0.1% value is 920).

BANDPASS FILTER – SPICE COMPARISON

(See page 53)

$$K := 10^3 \quad uF := 10^{-6} \quad R1 := 6.34 \cdot K \quad R2 := 80.6$$

$$R3 := 127 \cdot K \qquad\qquad C1 := 0.1 \cdot uF$$

$$C2 := C1 \quad BF := 400 \quad LF := 600 \qquad DF := 1$$

$$i := 1 .. \left(\frac{LF - BF}{DF}\right) + 1 \quad F_i := BF + DF \cdot (i-1) \quad s_i := 2 \cdot \pi \cdot F_i \cdot \sqrt{-1}$$

Transfer function:

$$G(R1, R2, R3, C1, C2, s) :=$$

$$\left| \frac{\dfrac{s_i}{R1 \cdot C1}}{(s_i)^2 + \dfrac{s_i}{R3} \cdot \left(\dfrac{1}{C1} + \dfrac{1}{C2}\right) + \dfrac{1}{R3 \cdot C1 \cdot C2}\left(\dfrac{1}{R1} + \dfrac{1}{R2}\right)} \right|$$

$$Tr := 0.02 \quad Tc := 0.1 \quad T := \begin{bmatrix} -Tr & -Tr & -Tr & -Tc & -Tc \\ Tr & Tr & Tr & Tc & Tc \end{bmatrix} \quad Nc := 5$$

$$p := 1 .. Nc \qquad dpf := 0.0001 \qquad Q := dpf \cdot identity(Nc) + 1$$

$$Vo_i := G(R1, R2, R3, C1, C2, s)$$

$$Vr_{i,p} := G(R1 \cdot Q_{p,1}, R2 \cdot Q_{p,2}, R3 \cdot Q_{p,3}, C1 \cdot Q_{p,4}, C2 \cdot Q_{p,5}, s)$$

$$Sen_{i,p} := \left(\frac{Vr_{i,p}}{Vo_i} - 1\right) \cdot \frac{1}{dpf} \qquad F_1 = 400 \qquad S1_{1,p} := Sen_{1,p}$$

$$F_{102} = 501 \qquad S1_{2,p} := Sen_{102,p}$$

$$S1 = \begin{bmatrix} -0.966 & 2.699 & 2.745 & 1.739 & 2.739 \\ -1.008 & -0.642 & 0.349 & -1.151 & -0.151 \end{bmatrix} \quad \begin{matrix} \text{Sensitivities at 400 Hz} \\ \text{Sensitivities at 501 Hz} \end{matrix}$$

$$L_{i,p} := if(S1_{1,p} > 0, 1 + T_{1,p}, 1 + T_{2,p})$$

$$H_{i,p} := if(S1_{2,p} > 0, 1 + T_{2,p}, 1 + T_{1,p})$$

$$Vev_{1,i} := G(R1 \cdot L_{i,1}, R2 \cdot L_{i,2}, R3 \cdot L_{i,3}, C1 \cdot L_{i,4}, C2 \cdot L_{i,5}, s)$$

$$Vev_{2,i} := G(R1 \cdot H_{i,1}, R2 \cdot H_{i,2}, R3 \cdot H_{i,3}, C1 \cdot H_{i,4}, C2 \cdot H_{i,5}, s)$$

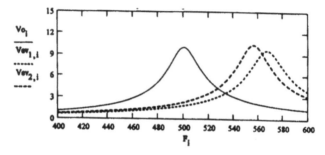

FIGURE 75. Duplicate of Spice probe plot.

MCA OF BPF

(See page 90)

$$K := 10^3 \qquad uF := 10^{-6} \qquad R1 := 6.34 \cdot K \quad R2 := 80.6$$

$$R3 := 127 \cdot K \qquad C1 := 0.1 \cdot uF \qquad C2 := C1 \qquad Bf := 400$$

$$LF := 600 \quad DF := 1 \quad lit := \frac{LF - BF}{DF} + 1 \qquad i := 1..lit$$

$$f_i := BF + DF \cdot (i - 1) \quad s_i := 2 \cdot \pi \cdot f_i \cdot \sqrt{-1} \quad N := 1000 \quad Nc := 5$$

$$k := 1 .. N \quad w := 1 .. Nc \quad z_w := rnorm(N, 0, 1)$$

$$Tr := 0.02 \quad Tc := 0.1 \quad T := \begin{bmatrix} -Tr & -Tr & -Tr & -Tc & -Tc \\ Tr & Tr & Tr & Tc & Tc \end{bmatrix}$$

$$Tn_{k,w} := \frac{(T_{2,w} - T_{1,w})}{6} \cdot \left[(z_w)_k + 3 \right] + T_{1,w} + 1$$

$$F1(R1, R2, R3, C1, C2, s) :=$$

$$\left| \frac{\dfrac{s_i}{R1 \cdot C1}}{(s_i)^2 + \dfrac{s_i}{R3} \cdot \left(\dfrac{1}{C1} + \dfrac{1}{C2} \right) + \dfrac{1}{R3 \cdot C1 \cdot C2} \left(\dfrac{1}{R1} + \dfrac{1}{R2} \right)} \right|$$

$$Vo_i := |F1(R1, R2, R3, C1, C2, s)|$$

MCA:

$$Vm_{k,i} := |F1(R1 \cdot Tn_{k,1}, R2 \cdot Tn_{k,2}, R3 \cdot Tn_{k,3}, C1 \cdot Tn_{k,4}, C2 \cdot Tn_{k,5}, s)|$$

$$Vmax_i := max(Vm^{<i>}) \quad Vpk := max(Vmax)$$

FMCA:

$$k := 1 .. 2^{Nc} \quad Re_{k,w} := k \quad Re_{k,w+1} := floor\left(\frac{Re_{k,w}}{2} \right)$$

$$Dr_{k,w} := Re_{k,w} - 2 \cdot Re_{k,w+1} \quad Tf_{k,w} := if(Dr_{k,w} = 0, 1 + T_{1,w}, 1 + T_{2,w})$$

$$Vf_{k,i} := |F1(R1 \cdot Tf_{k,1}, R2 \cdot Tf_{k,2}, R3 \cdot Tf_{k,3}, C1 \cdot Tf_{k,4}, C2 \cdot Tf_{k,5}, s)|$$

$$Vfhi_i := max(Vf^{<i>}) \quad Vfpk := max(Vfhi)$$

DERIVATION OF 400-HZ FWR DC EQUATION

(See page 143).

$$K := 10^3 \qquad R1 := 630 \qquad R2 := 14 \cdot K \qquad R3 := 20 \cdot K$$

$$R4 := 20 \cdot K \qquad R6 := 10 \cdot K \qquad R8 := 20 \cdot K \qquad R5 := 20 \cdot K$$

$$Vrms := 1 \qquad Vpk := 1 \qquad Vdc := 1 \qquad Vin := 0.25 \cdot Vrms$$

On pos half cycle of input:

$$Vo3p := \frac{Vin \cdot R2 \cdot R8 \cdot \sqrt{2}}{R1 \cdot R5} \qquad Vo3p = 7.857 \circ Vpk$$

On neg half cycle:

$$Vo3n := -Vo3p \qquad Vo3n = -7.857 \circ Vpk$$

Output via U2A:

$$Vo3 := \frac{Vin \cdot R2 \cdot R4 \cdot R8 \cdot \sqrt{2}}{R1 \cdot R3 \cdot R6} \qquad Vo3 = 15.713 \circ Vpk$$

Sum Vn2 and Vo2:

$$Vo3s := Vo3n + Vo3 \qquad Vo3s = 7.857 \circ Vpk$$

Dc average of output over one complete cycle:

$$\frac{Vo3p + Vo3n + Vo3}{\pi} = 5.002 \circ Vdc$$

$$\frac{Vo3p + Vo3n + Vo3}{\pi} = \frac{Vin \cdot \sqrt{2}}{\pi}$$

$$\cdot \left(\frac{R2 \cdot R8}{R1 \cdot R5} - \frac{R2 \cdot R8}{R1 \cdot R5} + \frac{R2 \cdot R4 \cdot R8}{R1 \cdot R3 \cdot R6} \right)$$

$$= \frac{Vin \cdot \sqrt{2}}{\pi} \cdot \left(\frac{R2 \cdot R4 \cdot R8}{R1 \cdot R3 \cdot R6} \right)$$

$$\frac{Vin \cdot \sqrt{2}}{\pi} \cdot \left(\frac{R2 \cdot R4 \cdot R8}{R1 \cdot R3 \cdot R6} \right) = 5.002 \circ Vdc$$

Note that R5 drops out. If R5 were not the correct value, the rectified half cycles would be of different amplitude, but the dc average would remain the same. Hence a scope should be used to test for correct ripple, thereby checking R5, R7, and C2.

References

1. *Introduction to Random Signal Analysis and Kalman Filtering*, R.G. Brown, Wiley, 1983, p. 44.
2. *Principles of Active Network Synthesis and Design*, G. Daryanani, Wiley, 1976.
3. *Tolerance Design of Electronic Circuits*, Spence and Soin, Addison-Wesley, 1988.
   Emphasis on MCA and circuit design to meet specifications with maximum manufacturing yield. Terminology: EVA = "Vertex Analysis;" RSS = "Method of Moments."
4. *Tolerance Design*, C. M. Creveling, Addison-Wesley, 1997.
   Mechanical and economic aspects of tolerance analysis.
5. *New Tolerance Analysis Methods*, R. Boyd, RF Design 97 Conference & Expo, Conference Papers, 1997, p. 75 and p. 307.
   Seminal paper.
6. Will your design keep on working? Divekar and Apte, *Test & Measurement World*, 3/89, p. 81.
   Excellent treatment of sensitivity.
7. *Spectral Analysis and Its Applications*, Jenkins and Watts, Holden-Day, 1968, p. 76.
   Definition of variance of a non-linear function.
8. *Surveying*, Moffitt and Bouchard, 6th edition, Harper & Row, 1975, pp. 166, 168.
   Derivation of RSS formula from total differential.
9. *An Introduction to Error Analysis*, 2nd edition, J.R. Taylor, University Science Books, 1997.
   Subtitle: The Study of Uncertainties in Physical Measurements.
10. MIL-STD-785B, Task 206, Electronic Parts/Circuits Tolerance Analysis.

11. *A Handbook of Active Filters*, Johnson, et al., Prentice-Hall, 1980, p. 105.
12. *Statistical Tolerancing*, Brady and Odorizzi, Electronic Design, 4/1/178, p. 134.
13. *Using Monte Carlo Simulations to Introduce Tolerance Design to Undergraduates*, Humann, Pierre, Legowski, and Long, IEEE Trans Education, Vol. 42, No. 1, Feb. 1999.
14. *State Space Averaging with a Pocket Calculator*, R. Boyd, High-Frequency Power Conversion Conference Proceedings, Santa Clara, CA, 1990, p. 283.
15. *Tolerance Analysis of Electronic Circuits Using MATLAB*, R. Boyd, CRC Press LLC, 1999.

T - #0543 - 101024 - C0 - 165/108/12 - PB - 9780849323393 - Gloss Lamination